CHEMISTRY

TODAY AND TOMORROW

The Central, Useful, and Creative Science

Ronald Breslow

American Chemical Society
Washington, DC

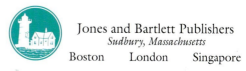
Jones and Bartlett Publishers
Sudbury, Massachusetts
Boston   London   Singapore

Library of Congress Cataloging-in-Publication Data

Breslow, Ronald.
    Chemistry today and tomorrow: the central, useful, and
    creative science / Ronald Breslow.
        p.   cm.
    Includes bibliographical references and index.
    ISBN 0-8412-3460-4 (alk. paper)
    1. Chemistry.   I. Title.
QD31.2.B75   1997
   540—dc21

96-45078

CIP

Figure Credits

1.3, Painting by Edward Allens Schmidt courtesy of Fisher Scientific Company;
2.1, Mallinckrodt Chemical Inc.; **2.2,** Robert Mach, North Carolina Baptist
Hospital, Bowman Grey School of Medicine, Wake Forest University; **2.3,** Vertex
Pharmaceuticals; **2.4,** Vistakon; **3.1,** Monsanto; **3.2,** Amoco; **3.3,** Lonza; **3.4,**
Amoco; **3.5,** NASA; **3.6,** Northrop Grumman/ USAF; **3.7,** Rollerblade; **3.8,**
Amoco; **3.10,** Zeneca; **4.3,** Eastman Chemical; **4.4,** General Motors **4.6,** NASA;
5.1, International Business Machines Corporation; **5.3,** Jimmy W. Crawford,
Research Triangle Institute; **5.4,** Molecular Simulations, Inc.; **6.3,** Eric Kool,
University of Rochester; **6.5,** Shell Research and Technology Centre, Amsterdam;
6.6, Chevron Chemical Company; **7.6,** AlliedSignal Specialty Chemicals; **8.1,**
Analytica of Branford, Inc.; **8.2,** Ariad Pharmaceuticals; **8.6,** National Research
Council.

Copyright © 1997 American Chemical Society

Educational, trade, bookclub and corporate customers should acquire this title
through the following Sales and Customer Service Office:
Jones and Bartlett Publishers
40 Tall Pine Drive
Sudbury, MA 01776
(800) 832-0034/ (508) 443-5000
info@jbpub.com
http://www.jbpub.com

PRINTED IN THE UNITED STATES OF AMERICA

ISBN 0-8412-3460-4 American Chemical Society Washington, DC
ISBN 0-7637-0463-6 Jones and Bartlett Publishers Sudbury, MA

Chemistry Today and Tomorrow

CONTENTS

DATE DUE

vi

PREFACE

Chemistry is one of the oldest sciences, and it has certainly been one of the most productive in improving human life. The industries that use chemistry to manufacture products—the chemical process industries—are the largest segment of manufacturing in advanced societies; in the United States, for example, they account for more than 30% of all manufacturing. This statistic does not include the output of related industries, such as electronics, automobiles, or agriculture, that use the products of the chemical process industries.

Much of what is done in designing and producing chemical products such as modern medicines is unknown to the average person or taken for granted. Most people are not aware that it is chemistry that creates such useful substances. We often hear about "toxic chemicals" or "chemical pollution" without hearing about the absolutely central role that chemistry plays in human well-being. In this book I attempt to present a more balanced picture.

I hope the audience will include students who might enter careers in science, or a field in which some knowledge of science plays a role. I also hope that adults will read the book to learn something about a field that affects all our lives and that holds tremendous promise for the future.

I do not assume that the readers have taken any course in chemistry, but believe that at some point they have learned that "chemicals" consist of molecules made up of atoms of the elements. For example, a molecule of water contains two atoms of hydrogen and one atom of oxygen, hence the formula H_2O. These atoms are linked together by chemical bonds. Bonds are commonly shown as lines connecting the atoms, so H_2O can be shown as H—O—H. This representation simply indicates that each hydrogen atom is linked to the oxygen atom. To understand this book, it is not necessary to know more about what the bonds consist of, or why they are able to link the atoms.

The book is organized so that the early part deals mainly with the practical contributions that chemistry makes, and will make in the future, to civilized life. The later part of the book tells a bit about how chemists make these contributions. Even this part is technically simple, but it is aimed at conveying the human activity behind this large field. For young people thinking about possible scientific careers, or their parents, this section may be particularly useful. As in all the chapters, it indicates a future in which new generations of chemists will be making the contributions.

It is difficult enough surveying the vast field of chemistry as it exists, and even more difficult making predictions about the future. To assist me with this task, I have asked a group of particularly far-seeing chemists to give me their own predictions of where the future lies. Some exciting ideas have come from these people (although of course they are not responsible if the future does not exactly correspond to the predictions). I acknowledge them here: Brian Bent, John Bercaw, Robert Bergman, Esther Breslow, Jean Chmielewski, Harry Gray, Kendall Houk, Martin Karplus, Stephen Lippard, Harden McConnell, Ann McDermott, Arthur Patchett, Kenneth Raymond, John Ross, Alanna Schepartz, Stuart Schreiber, Richard Schrock, Edward Solomon, Gabor Somorjai, Barry Trost, George Whitesides, Richard Zare, and Ahmed Zewail.

In addition, a number of people—chemists and others—have read this book at various stages and made useful suggestions. Jeff Holtmeier served as development editor. My resident editor, Esther Breslow, made many useful comments, as did my colleague Brian Bent and Madeleine Jacobs, the editor of *Chemical & Engineering News*. Other readers who were very helpful included Steve Baldwin, Frank Cardulla, Margaret Holland, Rita Nalebuff, and Pat Smith.

ABOUT THE AUTHOR

Ronald Breslow is Professor of Chemistry at Columbia University, and the President of the American Chemical Society. He first became interested in chemistry when he was a grade-school student, fascinated by the "magic" by which chemistry can transform one substance into another. After graduating from Harvard as a chemistry major in 1952, he earned a Master's degree in biochemistry and medical science in 1953 and then a Ph.D. in organic chemistry in 1955.

He joined the Columbia University chemistry faculty in 1956 and has pursued a teaching and research program in physical organic chemistry and bioorganic chemistry ever since. His research interests are in the synthesis of new molecules with interesting properties, particularly molecules that act as catalysts by imitating enzymes, and in understanding how chemical reactions occur. He has received teaching awards and many awards for his research, including the U.S. National Medal of Science awarded by President Bush in 1991. In addition to more than 300 research papers, he has published *Organic Reaction Mechanisms* and *Enzymes: The Machines of Life*.

Introduction to Chemistry: The Central, Useful, and Creative Science

> You see things; and you say, "Why?" But I dream things that never were; and I say, "Why not?"
>
> —George Bernard Shaw, *Back to Methuselah*, Act 1, Part 1

What is chemistry?

Chemistry is the science that tries to understand the properties of substances and the changes that substances undergo. It is concerned with substances that occur naturally—the minerals of the earth, the gases of the air, the water and salts of the seas, the chemicals found in living creatures—and also with new substances created by humans. It is concerned with natural changes—the burning of a tree that has been struck by lightning, the chemical changes that are central to life—and also with new transformations invented and created by chemists.

Chemistry has a very long history. In fact, human activity in chemistry goes back to prerecorded times.

What do chemists do?

As the quotation at the head of this chapter indicates, chemists are involved in two different types of activity. Some chemists investigate the natural world and try to understand it, while other chemists create new substances and new ways to perform chemical changes that do not occur in nature. Both activities have gone on since the first appearance of humans on earth, but the pace has increased enormously in the last century or so.

What was some of the earliest chemistry?

Curiosity about natural substances led to some of the earliest adventures in isolating pure chemical materials from nature. Humans discovered that they could extract the colors from flowers and some insects and use them to make pictures and to dye cloth. Only in the last century have chemists learned the detailed chemical structures of these natural colors. From earliest times humans have also been making new substances by performing chemical transformations. The first such new substances were probably soap and charcoal.

When wood is heated it loses water and produces char-coal. In this process the cellulose of wood—a chemical compound containing carbon, hydrogen, and oxygen all linked by chemical bonds—undergoes a chemical reaction that breaks the hydrogen and oxygen away as water and leaves the carbon behind as charcoal. A major chemical change has occurred—the process cannot be reversed to make cellulose again by just mixing the charcoal with water, since the oxygen and hydrogen atoms will not spontaneously form the needed bonds to carbon. Charcoal burns with a flame hotter than that of wood. Archaeological records indicate that charcoal has been used since prehistoric times.

Perhaps even earlier came the creation of soap. Soap is not a natural substance, but it can be made by heating fats with alkali to break some chemical bonds that link fatty acids to glycerin. Soaps are just the alkali salts of the resulting fatty acids. Since alkalis are formed when wood burns and are found in the ashes of wood fires, it is believed that the earliest humans noticed the unusual substances produced from fats that had dripped onto cooking fires.

These early "chemists" made such discoveries by accident, and for a long time accident was the principal means of discovery. Accident still remains important to discovery, but with our increasing chemical understanding we now usually create new chemical substances by design. How chemists create new molecules will be discussed further in Chapter 7.

After the early period of random discovery, humans began heating substances together intentionally to see what occurs. When a material that we now call iron ore was heated with charcoal, it produced iron metal, a new substance (we now use coke, produced from coal, instead of charcoal). Iron ore contains a chemical in which iron atoms are chemically bound to oxygen atoms. Heating it with charcoal lets the carbon atoms of charcoal bind to the oxygen atoms and carry them off as the gas carbon monoxide, leaving iron behind. Only gold and some metals related to platinum occur naturally as metals; all others are made from their ores by such chemical processes.

When copper and tin were heated together, the copper atoms and tin atoms linked up with metallic bonds, producing the alloy bronze, which is harder than either copper or tin. In the Bronze Age, starting at about 3600 B.C., the hardness of this metallic alloy made it the dominant material for tools and weapons. Bronze was the first metal that could hold a sharp edge.

Egyptians made glass as early as 1400 B.C. by heating some natural minerals together. Glass is formed when this heating causes major chemical changes (Figure 1.1).

Much of the rise of civilization has involved humans creating new substances by transforming natural ones to better meet their needs. Tanning hides to make leather, for example, changes their chemical nature. Even cooking foods alters their chemical structures. Every substance in the world is made up of "chemicals", either in the form of chemical compounds in which atoms are linked by chemical bonds or, in a few cases such as helium gas, as unlinked atoms. No substance can be called "chemical-free". Indeed, "natural" chemicals do not always have advantages. Some of the most dangerous poisons known are natural chemicals, produced by bacteria and other living things.

Modern chemistry is devoted to understanding the chemical structures and properties of natural chemicals and

3

FIGURE 1.1

Glass was one of the earliest chemical inventions, used to make this ancient vase.

of chemicals created by building on what nature has supplied. The remainder of this book will describe many of the contributions to everyday life resulting from this knowledge, and a sample of what is left to be done by the current and future generations of chemists.

Why do chemists call their discipline the "central science"?

Chemistry touches many other scientific fields. It makes major contributions to agriculture, electronics, biology, medicine, environmental science, computer science, engineering, geology, physics, metallurgy, and mineralogy, among many others. It does not ask the physicists' question: What is the ultimate nature of all matter? Instead it asks the chemists' questions: Why do the substances of the world differ in their properties? How can we control and most effectively utilize these properties?

Interesting and exciting as the physics question is, answers to the chemists' questions allow us to create new medicines, make new materials for shelter and clothing and transportation, invent new ways to improve and protect our food supply, and improve our lives in many other ways as well. Thus we see chemistry as "central" to the human effort to move above the brutish existence of our caveman ancestors into a world where we can exist not only in harmony with nature, but also in harmony with our own aspirations.

What makes chemistry the "useful science" and the "creative science"?

The two questions are linked. Some chemists explore the natural world and find useful chemical substances not known before. This exploration has been carried out extensively by examining the chemicals found in plants and animals on land, and it still goes on. Now there is also a major search for new chemicals from plants and animals in the seas (Figure 1.2). Once these chemicals are isolated and their chemical structures are determined, the creativity of chemists takes over.

FIGURE 1.2

Chemists are finding useful new chemicals, including medicines, in undersea organisms.

Normally we would not continue to harvest the living sources of useful new drugs, for instance—this could be too destructive and too costly. Instead chemists devise ways to synthesize the newly discovered compounds, to create them from other simpler materials, so they can be readily available. Sometimes the original chemical structures are altered by creative synthesis, to see whether the properties of a novel relative of the natural compound are even better.

There is a reason that the search for useful natural chemicals often pays off. The natural world is not the peaceful place we dream of—there are fierce battles for survival. Insects eat plants, and some plants have developed chemicals that will repel those insects. When we learn what those chemicals are, we can make them synthetically and use them to help protect our food plants. Bacteria can infect plants, animals, and other microorganisms such as yeasts and molds, not just humans. Some organisms have developed powerful antibiotics to protect themselves. Most of the effective antibiotics in human use have come from the exploration of nature's chemistry, although sometimes the medicines we use are versions improved by chemists.

Insects also use chemicals to signal each other for mating. As we learn what those chemicals are, we can make

FIGURE 1.3

A medieval alchemist in his laboratory. Heating various mixtures in the hope of producing gold, alchemists failed in their quest but made many interesting discoveries.

them and attempt to control the populations of harmful insects.

The most creative act in chemistry is the design and creation of new molecules. How is this done? This question will be more fully addressed in Chapters 2 and 7, but a brief answer is given here. New chemicals used to be made by what chemists irreverently refer to as "shake and bake": Heat up some mixture and see what happens, as in the earlier examples of making metals and glass. The alchemists of the past devoted themselves to heating up various mixtures in the vain hope to turn lead into gold. They did not succeed, but they did create some interesting new chemical processes and new substances (Figure 1.3).

Syntheses are now normally designed using the fundamental principles that chemists have discovered. As many as 30 or more predicted chemical steps are sometimes needed, in a sequence, to permit the synthesis of a complicated molecule from available simple chemicals. This could not be done without a clear understanding of chemical principles.

What are some fundamental principles of chemistry?

The first and most important principle is that chemical substances are made up of **molecules** in which atoms of various elements are linked in well-defined ways. The second principle is that there are somewhat more than 100 elements, which are listed in the **periodic table** of the elements. The third principle is that those elements, arranged according to increasing numbers of protons in their nuclei, have **periodic properties**. That is, as the elements increase in their atomic number (number of protons in the nucleus), every so often an element appears that is similar in its properties to one that has occurred earlier in the table.

For example, after **lithium**, with three protons in the nucleus, come other elements whose properties are decreasingly like those of lithium until suddenly **sodium** appears, with a nucleus containing 11 protons. Sodium is quite similar to lithium in many respects. The arrangement of the periodic table puts such similar elements below each other in columns. Chemical reactions that lithium undergoes will also occur with sodium, although not with the same speed or energy; lithium and sodium are similar, but not identical.

Another principle is that the ways in which atoms are linked strongly affects the properties of chemical substances. This is particularly evident when covalent links (bonds) are involved. **Covalent bonds**, in which two atoms are held together by a pair of electrons shared between them, are the bonds that hold the atoms of carbon, oxygen, and hydrogen together in cellulose, for instance. Most covalent bonds do not break easily, which is why intense heating is needed to turn cellulose into charcoal. The precise arrangement of the links determines chemical properties. By contrast, a salt such as sodium chloride has what are called **ionic bonds**. The sodium and the chlorine are not directly linked, just held together by the attraction of the positive sodium ion for the negative chloride ion. When sodium chloride is dissolved in water, the sodium ion and the chloride ion drift apart.

For example, **ethyl alcohol** (the "kick" in alcoholic beverages) has the chemical formula C_2H_6O. That is, a molecule of ethyl alcohol contains two carbon atoms, six hydrogen atoms, and one oxygen atom. However, there is

7

another chemical called dimethyl ether that has the same formula, but in which the atoms are linked differently. Because of this, the two chemicals have very different properties. Chemists call such compounds—with the same overall formula but different atomic arrangements—isomers of each other. A chemist's symbolic drawing of the two isomers with formula C_2H_6O is shown in Figure 1.4.

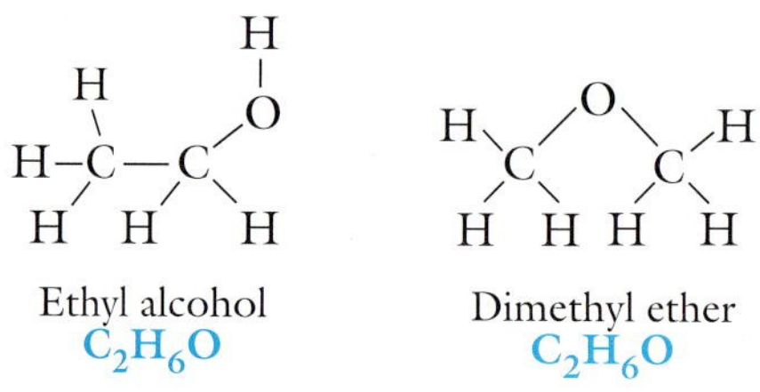

There is another much more subtle difference among chemical structures that has to do with the three-dimensional arrangement of atoms in space. Two chemicals can differ, even when all the same atomic linkages are present, if the spatial arrangements are different. Differences in spatial arrangement can have several aspects, but the most interesting has to do with handedness, or what chemists call chirality. When a carbon atom carries four different chemical groupings, there are two different ways they can be arranged. For example, in the amino acid alanine the central carbon atom carries four different groups: a hydrogen atom, a nitrogen atom, and two carbon atoms that differ in what is attached to them. Thus they can be arranged in two different ways, as Figure 1.5 shows. These arrangements differ in the way that a right hand

FIGURE 1.4

The molecules of ethyl alcohol (*left*) and of dimethyl ether (*right*) have the same atomic composition, but the atoms are arranged differently. The two chemicals have very different properties.

8

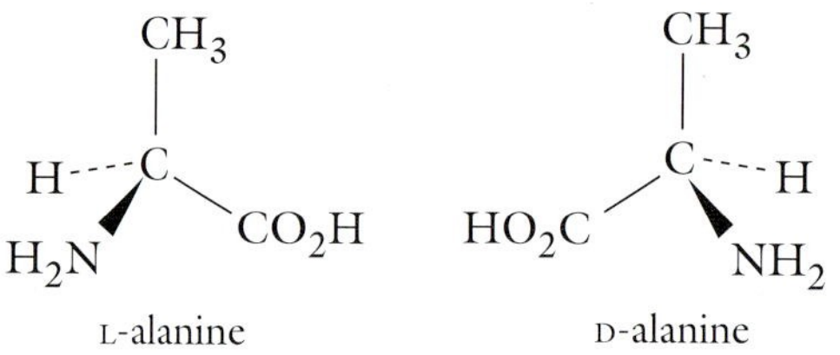

FIGURE 1.5

Two isomers of the amino acid alanine that differ in the way a right hand and a left hand differ: they are mirror images of each other. L-Alanine (*left*) is part of all human proteins. D-Alanine (*right*) is produced by bacteria and incorporated in their cell walls. Some antibiotics take advantage of this difference by binding to the D-alanine in bacteria but ignoring the L-alanine in human cells.

and a left hand differ, or a right shoe and a left shoe—that is, they are mirror images of each other. The one labeled L-alanine is commonly found as part of our proteins, but the mirror image D-alanine is not. This handedness of natural proteins means that they interact differently with right-handed and left-handed molecules, just as a right shoe fits one foot but not the other. Many medicinal chemicals can exist in two mirror image forms, and it is usually true that one of them is the better medicine and the other may even be harmful (*see* Chapter 2). One of the challenges that modern chemists are addressing is finding good ways to make new molecules with the desired handedness (*see* Chapter 7).

There are some fundamental principles governing chemical reactions, by which molecules change into other molecules. One principle is that such changes will not occur if the products of the reactions would be much less stable (have higher energy) than the starting materials. Just as rocks roll downhill but not uphill, chemical reactions spontaneously roll "downhill" to lower energy states. (Energy is not the only consideration, since chemical reactions also go in directions to maximize disorder, which is called entropy by chemists. A simple analogy is that shuffling a new deck of cards tends to put them into random arrangements, and further shuffling does not put them back into order again.)

Another principle is that even favorable reactions, whose products are lower in energy or more disordered than the starting materials, do not necessarily occur rapidly. This is a good thing—the burning up of all living things by reaction with the oxygen of the air is a favorable process energetically, but luckily it does not happen readily unless the temperature is very high, as in a flame.

This final example reflects a related principle: even reactions that end up with low-energy products need some extra energy to pass through intermediate stages that are not stable. A good analogy would be taking a trip from Denver to San Francisco. The overall trip is downhill from Denver at 5000 feet to San Francisco at sea level, but extra energy is needed to get over the mountain ranges along the way. The extra energy for chemistry—needed to climb such "mountain ranges" along the way—is available at high temperature, so most chemical reactions speed up when they are heated. Catalysts find another way to speed up favorable reactions (*see* Chapter 6).

How large a field is chemistry?

Virtually every human activity uses some of the materials supplied or processed by chemistry, and many such activities involve chemical changes as well. This is the subject of the rest of this book, but here is a brief outline.

In 1995, more than one million new chemicals were created worldwide by chemists. The pace is accelerating. Some of the new molecules are examined as possible new medicines (*see* Chapter 2), and others are used to make new materials such as useful plastics (*see* Chapter 3). Newly invented chemical reactions are often applied to manufacture pharmaceutical compounds and other chemicals more efficiently.

Biochemistry and molecular biology are on the borderline with medicine and biology. Here scientists work to unravel the chemistry of life and of disease. With this information, medicinal chemists can design new medicines (*see* Chapter 2).

Other chemists work on the environment, studying what is happening (e.g., the chemical reactions that have caused the creation of the ozone hole) and learning how to make manufacturing processes environmentally friendly (Chapter 4). Developing such processes is a major activity in the chemical and pharmaceutical industries, which have pledged to see that modern manufacturing fully respects the environment as part of "green chemistry". For example, a new chemical plant in Tennessee returns water to the river that is cleaner than the water it takes in.

As the discussion in Chapter 5 indicates, chemistry contributes the special materials that make modern electronics possible, but it also uses computers extensively. The field of computational chemistry uses modern chemical theory to predict (1) the properties of unknown chemicals, (2) the geometric shapes that unknown chemicals will have, (3) the reactions that will occur between molecules that have not yet been examined, (4) the speed of those unknown reactions, and (5) synthetic sequences by which complex new molecules can be made efficiently.

Catalysis is at the heart of the chemistry of life, and also of modern chemical manufacturing. This special important part of chemistry is described in Chapter 6. Many chemists are involved in the creative synthesis of new molecules. Chapter 7 describes how this is done.

10

Some chemists devote themselves to determining the detailed structures of newly discovered or newly created chemicals. Other experimental chemists study chemical reactions in exquisite detail, using very modern techniques, to learn the exact geometry of the reactions, the ways in which extra energy is used to make the reaction occur rapidly, and how the energy ends up in the products. The results feed into computer models that are used to predict unknown processes, helping chemists to invent new reactions. Both these activities are described in Chapter 8.

Chemical engineering is concerned with turning laboratory chemistry into practical manufacturing methods. Chemical engineers help design new chemical plants and help redesign existing ones for better efficiency and environmental safety.

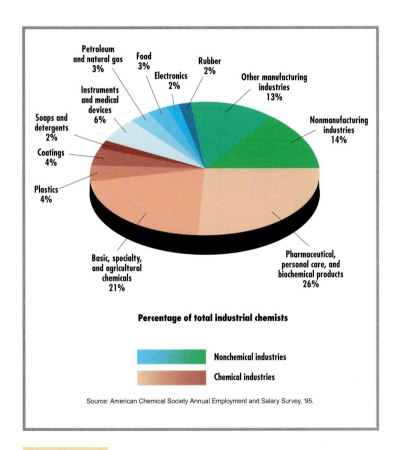

11

FIGURE 1.6

Industrial chemists work in many areas outside the chemical industry.

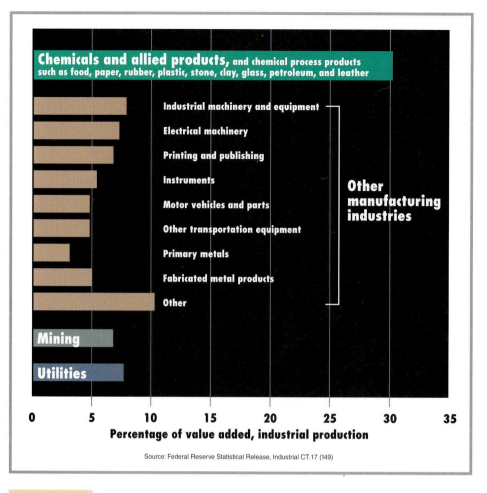

FIGURE 1.7

The chemical process industries, in which products are made by performing chemical transformations, include more than 30% of all industrial output in the United States. Because primary metals, such as iron and steel, are also produced by chemical processes designed and monitored by chemists, they should also be included. Most people don't realize that these industries amount to such a large segment of the economy.

The economic benefit from all this activity is very large. In addition to the chemical industry itself and the closely related pharmaceutical industry, we must also include the oil industry, which converts crude oil into gasoline and other products by changing some chemical structures; the rubber industry; the paper industry; the glass and tile industry; the steel industry; the food and beverage industry; the textile industry; the leather industry—all use chem-

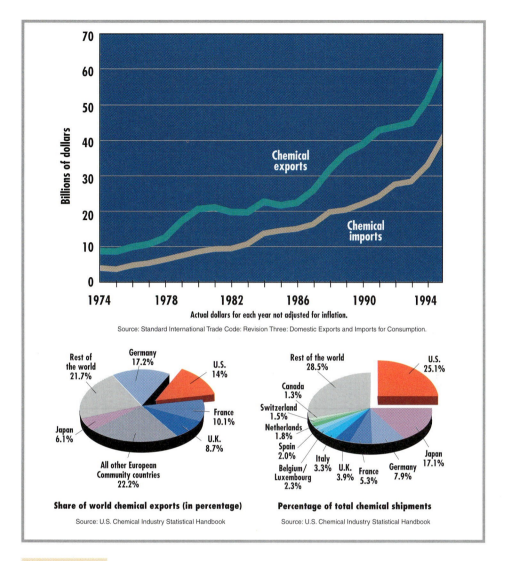

Share of world chemical exports (in percentage)

Percentage of total chemical shipments

FIGURE 1.8

U.S. share of worldwide chemical output.

istry extensively. They also employ chemists; about 50% of industrial chemists work in these related industries rather than in "chemical" companies (Figure 1.6).

All together these are called the **chemical process industries**, in which chemical reactions and procedures are used to make their products. They account for more than 30% of the total industrial production in the United States, in terms of value added (i.e., the economic value that their activities add after correcting for the value of their starting materials). No other manufacturing industry is as large, or even comes close (Figure 1.7).

The chemical industry alone, narrowly defined, employs one million workers, puts more money into research and development than any other industry, accounts for 10% of all U.S. exports, and is one of the few industries that has maintained a healthy balance of trade. International trade in chemicals is $300 billion annually, and the United States has 14% of the market (Figure 1.8).

The range of chemistry is also revealed in the divisions of the American Chemical Society, which collectively reflect the work of the 150,000 members of the organization. These divisions regularly meet to discuss progress and hear about research advances in their areas.

DIVISIONS OF THE AMERICAN CHEMICAL SOCIETY

Agricultural and Food Chemistry	Fertilizer and Soil Chemistry
Agrochemicals	Fluorine Chemistry
Analytical Chemistry	Fuel Chemistry
Biochemical Technology	Geochemistry
Biological Chemistry	History of Chemistry
Business Development and Management	Industrial and Engineering Chemistry
Carbohydrate Chemistry	Inorganic Chemistry
Cellulose, Paper, and Textile	Medicinal Chemistry
Chemical Education	Nuclear Chemistry and Technology
Chemical Health and Safety	Organic Chemistry
Chemical Information	Petroleum Chemistry
Chemical Technicians	Physical Chemistry
Chemical Toxicology	Polymer Chemistry
Chemistry and the Law	Polymeric Materials: Science and Engineering
Colloid and Surface Chemistry	Professional Relations
Computers in Chemistry	Rubber
Environmental Chemistry	Small Chemical Businesses

Human needs for new chemistry are as great as ever. We need and will create new drugs to fight diseases such as cancer, acquired immunodeficiency syndrome (AIDS), Alzheimer's disease, heart disease, and stroke that shorten our lives or diminish their quality. We will invent new ways to generate and store energy. Methods to isolate and concentrate the radioactive products from nuclear reactors will make nuclear energy much more acceptable. Methods to store electricity with new batteries will make it possible to convert to electric vehicles that are superior to and cleaner than the current gasoline models. New manufacturing processes will help us make the materials we need while protecting our environment.

We will improve computational chemistry to the point at which we can predict which new molecule to make for some desired properties, and determine how to make it. We will move from studying the properties of isolated molecules to fully understanding the properties of organized chemical systems, as in a living cell. We will learn how to make catalysts for our own needs that equal or exceed the natural enzymes in their effectiveness and selectivity. This will make it possible to carry out chemical manufacturing without using energy for high temperatures to speed up reactions, and without making the unwanted side products that result when reactions are not sufficiently selective.

In the succeeding chapters, I describe these and other aspects of our chemical future in more detail. This future depends on participation by future generations of chemists, the students of today, and support by the rest of society. However, there is one prediction we can make with assurance: some wonderful new things will be created or discovered that we have not even anticipated. Science, including chemistry, is constantly surprising us, and this will surely continue. As the physicist Leo Szilard said, "Prediction is difficult, especially of the future."

15

The Chemistry of Health and Life

Health and intellect are the two blessings of life.

—Menander of Greece, ca. 300 B.C.

Some of the most important contributions of chemistry and chemists have been to human and animal health. Improving health is the primary concern of the pharmaceutical industry, in which many chemists are employed creating new medicines (Figure 2.1). Many other chemical activities also contribute to human and animal health. For example, chemists have created insecticides that play a major role in decreasing the problems of malaria and other insect-borne diseases, especially in tropical countries. Sunscreens are used to help prevent skin cancer and painful sunburn. Artificial sweeteners are important for many people who should not use sugar.

Chemistry also plays a central role in the diagnosis of disease (Figure 2.2). X-ray film is a specialized product of the photographic industry, in which chemists design new films and new photographic processes. Magnetic resonance imaging (MRI) was invented as the result of studies on nuclear magnetic resonance spectroscopy applied to chemical problems. Tests of blood or urine that are part of most physical exams were invented by clinical chemists and are performed in analytical laboratories by other chemists.

In this chapter we will focus on two general topics: medicinal chemistry, by which new medicines are created, and biochemistry, by which the chemicals and chemical transformations of life are discovered.

What has medicinal chemistry contributed to human health in the past?

The average life expectancy of a male born in America in 1900 was only 47 years, but for a male born today in this country it is about 75 years. This incredible improvement can be credited in large part to contributions made by medicinal chemistry. Most important was the development of antibacterial agents.

Prior to the 1920s a bacterial infection was frequently fatal. Then chemists began to synthesize many new chemicals to serve as dyes for cloth, including some chemicals that had what is called a sulfonamide group. The German scientist Gerhard Domagk tested various of the newly made chemicals to see if any of them could kill bacteria, and in 1932 he tested a red-brown dye called Prontosil. It was effective in curing mice with otherwise fatal bacterial infections. He then tested it in a child with a hopeless case of bacterial blood poisoning, and she recovered.

FIGURE 2.1

Modern research chemists design and synthesize more than one million new chemical molecules each year, many to be examined as possible medicines.

18

Building on this lead—which was recognized with the 1939 Nobel Prize in Physiology and Medicine to Domagk—chemists created many new so-called sulfa drugs containing sulfonamide groups. Sulfanilamide was widely used, and it is still in use to some extent. However, with the discovery that chemicals could be found that would kill bacteria but not harm humans or animals, a major new research area opened, and many improved antibacterial chemicals were created or discovered. They are still the subject of active research, as will be described later.

Biochemical studies have made it clear why sulfanilamide can kill bacteria while not being toxic to humans: our biochemistry is different. Bacteria make the essential vitamin called folic acid, and sulfanilamide blocks the enzymes that make folic acid in bacteria; without it they cannot survive. Humans don't have those enzymes, so we take folic acid in as part of our diets. In a sense we are saved by a flaw in our biochemistry—our inability to make an important vitamin (but our ability to get it in our food).

The process by which sulfa drugs were discovered is called random screening. It is random because there is no logical reason to suspect that a clothing dye would be a selective killer of bacteria. Random screening is still used extensively today. However, with better information about biochemistry, medicinal chemists also increasingly use a process called rational drug design, in which particular molecules are designed and created for desirable biological

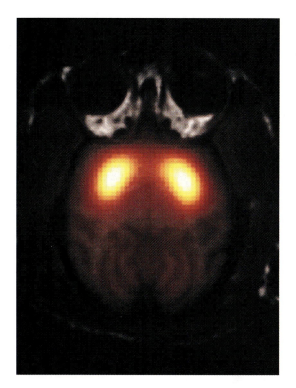

FIGURE 2.2

An image of a monkey brain created by positron emission tomography (PET), using synthetic radioactive chemicals, and magnetic resonance imaging (MRI). Such modern techniques have made the diagnosis of disease much easier.

effects. Surprises still occur—some chemicals designed to act as rational drugs for particular medical problems turn out, from random screening, to be "irrational" drugs for something else.

The discovery of sulfa drugs stimulated the modern field of antibiotics, but it was not the first contribution of chemistry to health. Disinfectants for surface wounds, such as iodine or phenol, came earlier, and so did anesthetics. Alcohol was one of the earliest anesthetics; the patient got drunk before a painful procedure. Later it was discovered that ether was more effective and made painless surgery and dentistry possible. Since that time better anesthetics have been created, including local anesthetics such as Novocain. Without anesthetics, modern surgery would not be possible.

By now medicinal chemicals are available to treat essentially every human disease, although for many important diseases the treatments are not fully effective yet. Besides antibacterials, we now have antivirals and antifungals, and treatments for parasites. However, improved antivirals and antifungals are still needed. We have medicines to treat stroke and heart attack, and others to treat ulcers. We have pain killers and antidepressants, and compounds to treat various hormone deficiencies. There are antihistamines, drugs to control motion sickness, and drugs to lower cholesterol levels. Anticancer and anti-AIDS drugs also exist, but these diseases are not yet under control. As we will see in the final section of this chapter, a lot of current work aims at overcoming them.

How do medicinal chemists design or discover medicines ?

The main procedure has been, until recently, random screening. This is still in use, but it includes an exciting new approach called combinatorial chemistry, by which literally thousands of new chemicals can be made at one time carrying chemical "tags" that indicate their identity. An entire group of chemicals is screened simultaneously against some biological target, such as an enzyme, and the effective chemicals are picked out and identified. The chemical structures of these effective molecules are taken as leads, and new related chemicals are then synthesized for testing. The chance that any one new chemical will show exciting biological activity in a random screen is small, but with the ability to make and screen thousands of chemicals at one time chemists have greatly increased the odds of finding a valuable drug.

Another procedure involves exploring the natural world for effective chemicals. Plants, fungi, insects, and bacteria themselves make many unusual compounds, and these are screened for useful properties. However, there is some rationality in this exploration as well. Bacteria attack animals, plants, and even fungi—we are not their only victims. Many organisms have developed ways to protect themselves. Some of the most effective antibacterial agents are products of this war of the microorganisms, with bacteria on one side and molds and fungi on the other. Penicillin, erythromycin, and vancomycin are examples of the defenses that molds create. We may continue to prepare them by growing the mold, but frequently chemists devise a better procedure to prepare them using synthetic chemistry.

In modern pharmaceutical companies about half of the research scientists are chemists, whose main activity is synthesizing new molecules or devising good ways to manufacture the ones that turn out to be useful. Increasingly they decide what to make by considering what is known about biological chemistry.

For example, we now know that **blood pressure** is regulated by hormones produced by particular enzymes. Furthermore, we also have learned how those enzymes work (*see* Chapter 6 for more on this topic). Medicinal chemists take two approaches to trying to block those enzymes, so as to lower blood pressure.

In one approach, they consider what the biological molecule is that normally binds to the enzyme. Then they make new molecules that look like the natural one but that are expected to bind more strongly. One of the important enzymes in regulating blood pressure has a zinc atom as part of it, so the new molecules are designed with chemical groups that are known to bind strongly to zinc. This strategy has been quite effective, and some of the drugs designed in this way are currently in medical use to treat high blood pressure.

The second approach involves learning more about the enzyme itself. Biochemists isolate and purify enzymes, and prepare them as crystals. Chemists then use X-rays to determine the three-dimensional structure of the enzyme, which

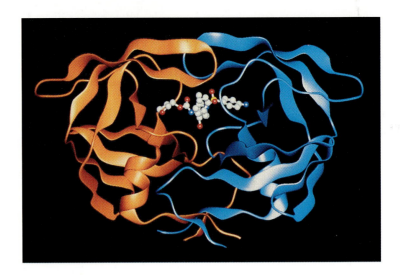

FIGURE 2.3

A computer representation of a potential drug binding into an enzyme that is vital for the functioning of HIV, the virus that causes AIDS. The protein is shown in a ribbon diagram, which depicts how the chains are folded but does not indicate the chemical details of the molecule.

22

they display on a computer screen. Then medicinal chemists make computer models of various possible drugs, and see whether they can fit them onto the enzyme structure shown in the computer (Figure 2.3). Often they can calculate how strongly the potential drugs will bind. (Computational chemistry is discussed further in Chapter 5.) Finally, the medicinal chemist selects a drug candidate to be made. Then chemical synthesis of the potential drug is done (*see* Chapter 7), and the compound is tested for its biological activity.

Although these approaches have been described for enzymes, related approaches are also used for hormones. That is, molecules are made that will interact with the body where the hormone would normally exert its effect. Usually there is not much information about the chemical structure of the place where the hormone acts, so the major approach is to imitate the chemical structure of the hormone itself. The imitator chemical can be an "agonist" that has the same effect as the hormone, if more such effect is needed, or an "antagonist" that lowers excessive hormone effects if it gets in the way and prevents the hormone from reaching the target.

What problems are medicinal chemists addressing now?

Good scientists have the ability to define, and then solve, important problems. Each pharmaceutical company has its own vision of what the most important research areas are. However, some have captured everyone's attention.

The resistance of some bacteria to our most effective antibiotics is a growing threat. As normal antibiotics kill off most of the bacteria that cause infections, a few bacteria survive that have changed their biochemistry in such a way as to be immune to current antibiotics. We could characterize this as "survival of the fittest". Since the survival of dangerous bacteria casts our own survival in doubt, we cannot ignore the problem. We have to be even fitter, by creating new antibiotics that the new strains of bacteria cannot evade.

For example, penicillin is very effective in killing bacteria, but some of them have developed enzymes that can destroy penicillin. Work is going on to develop variations of the penicillin chemical structure that will not be destroyed by bacterial enzymes; other work is going on to find drugs that will block those bacterial enzymes, and permit penicillin itself to function.

A particularly difficult problem is posed by the human immunodeficiency virus (HIV), which causes AIDS. This virus causes a cell to produce a special enzyme that performs a critical step in the viral life cycle. Medicinal chemists have designed drugs that block that enzyme, thus stopping the infection. Rational drug design was used, based on chemical knowledge about the enzymatic reaction. However, the virus can change its biochemistry rapidly, and new forms appear that are no longer inactivated by these drugs. A major battle is under way between this dangerous virus and an army of medicinal chemists, in many academic and industrial laboratories. It is not yet clear that we will win, or how.

HIV is not our only virus enemy. The viruses that cause influenza are a continuing threat, and there are still few good treatments for viral infections. Many pharmaceutical companies have general programs to find antiviral drugs.

Almost every drug company has an anticancer program. Several approaches look promising. In one, chemists are trying to devise, or discover, new chemicals that will kill

23

a cancer cell but not a normal one. This is a tough challenge, but some effective drugs have been created and more are coming. In another approach, chemists are developing drugs that will change the behavior of a cancer cell, making it act in a normal way. Some good leads are also available here. The problem is by no means solved, but not for lack of attention by chemists, biochemists, and biologists.

Another area of considerable current interest relates to organ transplants, such as transplanting a healthy heart into a diseased patient. The problem here is mainly that the body recognizes such a transplanted heart as foreign and tries to reject it. There is very active research on new medicinal chemicals that can block this rejection, so the transplant can be successful.

As medicinal chemistry has prolonged our life span, we can try to deal with some of the problems that come with old age. One of these is Alzheimer's disease, which destroys the quality of life of many older people. Pharmaceutical companies, and academic laboratories, are trying to develop drugs to deal with this challenge.

Drug delivery is a special area of medicinal chemistry. Modern plastics are being used to make devices and implants that will slowly release needed drugs or hormones in a controlled fashion. Even the pills we swallow, containing medicines or vitamins, are chemically designed to release their contents where needed—for instance in the stomach, rather than in the mouth. There is a lot of interest in learning how to steer drugs to their desired targets, especially to a growing cancer. Such steering would decrease the undesirable side effects that drugs could have on other parts of the body.

What have chemists contributed to understanding the nature of life?

The branch of chemistry concerned with understanding life is called biochemistry. There is also a field called molecular biology that is part of biochemistry but heavily focused on DNA and RNA, the molecules that carry and transfer genetic information. Some of the most exciting discoveries in modern science are being made in this rapidly changing field.

The food we eat is converted to all the chemicals—proteins, carbohydrates, fats, and hormones—that we need for life. Some special molecules are also made that supply the energy for the chemical changes that are part of life, such as those that power muscle movement. During this century biochemists have discovered essentially all the hundreds of chemical pathways that are involved.

Furthermore, all these biochemical changes are catalyzed by enzymes, proteins that make the chemical reactions faster and more selective. Biochemists have identified most of the enzymes involved, about 7000, and have learned much about how they work. More about enzymes and catalysis will be discussed in Chapter 6.

Enzymes and all other important proteins are made under the direction of genes, which carry the code to specify which amino acids will be used to make the protein, and in what sequence in a protein chain. Those chains of linked amino acids then fold into the correct shape to perform their function, as is discussed in Chapter 5. Biochemists and molecular biologists use new chemical methods to determine the molecular structures of the genes, and other chemical methods to change those structures so as to make different genes. This area holds great promise for future achievements in the application of chemistry to health problems.

Great progress has also been made in learning about the differences in chemistry that we find in different living creatures. For example, knowing some of the biochemistry that is important to bacteria but not to humans gives us a real advantage in designing new antibacterial agents. Also, plants have one special biochemical path that we now understand in much detail—photosynthesis, by which the energy of sunlight is used to power chemical synthesis.

I pointed out in Chapter 1 that chemical reactions will not occur if the products are much less stable than the starting materials. An exception to this rule occurs if extra energy can be supplied in some way, as it is in photosynthesis. This extra energy permits plants to convert carbon dioxide and water into oxygen and sugars, a process in which the products are much less stable than the starting materials. The energy needed to drive this otherwise unfavorable process comes from light.

25

In early work chemists learned the detailed molecular structure of chlorophyll, the chemical that is central to photosynthesis. In very recent work, recognized with the Nobel Prize in Chemistry in 1988, Harmut Michel, Johann Diesenhofer, and Robert Huber used a technique called X-ray crystallography to establish the chemical structure of the photosynthetic center, in which several proteins and chlorophyll molecules are organized. In this structure the incoming light is captured and used to make some energetic molecules.

The light energy in a chlorophyll molecule raises the energy of a chlorophyll electron so the electron can transfer to another molecule in the photosynthetic center. In the course of electron transfers some inorganic phosphate molecules become incorporated into a high energy molecule called ATP, the source of the chemical energy that drives many biochemical processes. For instance, ATP is the fuel that makes muscles move. Furthermore, the electron that is transferred ends up in a coenzyme derived from niacin (see Figure 7.2) and is used to carry out other biochemistry. The chlorophyll molecules that have lost electrons eventually get electrons back from water, converting the water to oxygen (O_2).

The organic chemist Melvin Calvin received the Nobel Prize in Chemistry in 1961 for showing the sequence by which photosynthesis converts carbon dioxide into sugars and other important chemicals such as amino acids. The idea of his central experiment was simple but brilliant, as many great scientific research ideas are.

It was known that carbon dioxide "labeled" with radioactive carbon-14 could be used in photosynthesis, and then the products would contain the carbon-14 and be radioactive. Calvin realized how he might learn the sequence by which these biochemical products were made.

He set up an apparatus in which algae were carrying out photosynthesis using ordinary carbon dioxide, water, and light, and he suddenly introduced some radioactive carbon dioxide into the apparatus. At a later time he dropped the whole mixture into boiling alcohol. This action stopped all the biochemistry, since the critical enzymes are not stable under such conditions.

If he waited only a short time for the boiling alcohol treatment after adding the carbon-14 carbon dioxide, and then isolated the chemical products, all the radioactivity was found in a simple three-carbon chemical related to the simplest sugars. If he waited longer before causing the alcohol to quench the reactions, radioactivity was still found in the three-carbon chemical but it was also found in chemicals related to glucose and fructose, which are six-carbon sugars. With still longer time lags before quenching, other biological chemicals became radioactive.

The first chemical that was labeled was the first one to incorporate carbon from carbon dioxide, while those that came later were the products of later biochemical reactions along a pathway. With this simple experiment the sequence was determined. Other work has identified all the enzymes that catalyze the changes involved.

What is the role of chemistry in molecular biology and biotechnology?

Modern techniques permit scientists to isolate and identify genes that specify the production of particular proteins, and then to incorporate those genes into organisms such as yeast. When the yeast is grown, it will produce the desired protein. Sometimes this natural protein is of importance in itself, as with insulin, for instance, or human growth hormone. Sometimes the scheme is modified by chemically altering the gene so that the novel protein produced has improved properties.

Chemists, including biochemists and molecular biologists, devise the methods to perform these operations and to isolate and purify the resulting proteins. This is an area of rapid and exciting growth, and much is expected of its future.

One special area has captured particular attention recently—the sequencing of the human genome. The genome is the entire mass of genetic DNA, in the nuclei of cells, that carries thousands of individual genes, each one consisting of several hundred or more bits of DNA acting as code letters. DNA is a large molecule made up of individual units that are linked by phosphate groups. Each unit contains a sugar, deoxyribose, carrying one of four bases that are symbolized by A, G, C, and T. They are copied into messenger RNA containing ribonucleotides—just like DNA but instead containing the sugar ribose—with the code letters A, G, C, and U (the T in DNA becomes a chemically related U in RNA), and this RNA then directs the synthesis of particular proteins.

The human genome has been estimated to contain several billion such units. About 10% of them are the code let-

27

ters for amino acids that are parts of proteins. The rest of the genome does not carry the code for proteins, and the function of this other DNA is not fully understood.

In this monumental project, sequencing the human genome, the chemical sequence in each gene will be identified, one code letter at a time. It is hoped that eventually we will understand the function of each gene, perhaps finding new proteins coded by sections of the genome we didn't know about.

A new continent is being explored, and no one knows for sure what will result. However, the chemical sequence in the genome determines what we become, and it has to be known before we can assert what mysteries this knowledge will solve.

Considering the importance of chemistry for understanding life, and the importance of medicinal chemistry for health, it is no surprise that the normal education of physicians involves a lot of chemistry. In their premedical studies in the United States they normally take at least two years of college chemistry courses, and often a biochemistry course as well. Then in medical school they take courses in biochemistry and in pharmacology that build on their chemical backgrounds. It is a big change from the education of doctors at the beginning of the century, when so little biological chemistry was understood and so few medicines were available. Modern doctors need to prepare for a future in which chemistry plays an even larger role in human health.

T H E   F U T U R E

Medicinal chemists agree that we are likely to develop effective treatments for the major lethal diseases of humankind. Very promising approaches to cancer therapy are under way. Eventual triumphs over at least some forms of this disease are expected; the major question is when they will be achieved. Also, strokes and heart diseases related to high cholesterol levels will be controlled, in part by healthier life styles and in part by new pharmaceutical agents. The battle against AIDS is showing progress, and more progress is expected.

The quality of life is also important, and many nonfatal diseases are targeted currently. As more is understood about the biology and chemistry of Alzheimer's disease, we can expect the development of effective pharmaceuticals. With aging the levels of some hormones decrease, and drugs that prevent this decrease or replace the hormones can lead to a more vigorous old age. Osteoporosis—the loss of bone that causes problems for older people, particularly women—is the target of pharmaceutical research now. Drugs to treat obesity will also add both life span and quality of life. Drugs that will assist in nerve regeneration after a stroke, drugs to treat schizophrenia, better treatments for arthritis—all are likely in the foreseeable future.

Advances in biochemistry trigger research in new medicinal areas. For instance, it has recently been found that the simple molecule NO, with one nitrogen atom bonded to one oxygen atom, is an important messenger molecule. It is formed in a biochemical reaction of the amino acid arginine, and stimulates many biological responses in humans. It helps regulate blood pressure. Medicinal chemists are now working on ways to control the levels of NO in the body, by controlling the activity of the enzyme that produces NO.

One of the most exciting new areas is gene therapy. Some human diseases are the result of defects in our own genes, not the attack of a microorganism. Medicinal chemists are trying to develop methods to deliver pieces of DNA to cells so they can be incorporated to replace the defective pieces. It is a challenging area, but progress is being made. In the future, success in gene therapy will help us treat many health problems not currently solvable.

Diabetes is a disease in which the body is unable to control the level of the sugar glucose in the blood. Normally the pancreas responds to high glucose levels by secreting insulin, but in one form of diabetes this control mechanism is defective. An exciting approach to treatment would be an "artificial pancreas" that could detect glucose levels and then add the correct amount of insulin to the blood to restore glucose to healthy levels. Glucose levels can be determined by electrochemistry using enzymes specific for glucose; the incorporation of a glucose sensor into a device with a minipump for delivering insulin is an attractive possibility for the future.

Much research is under way to develop better detection methods for diagnosis of diseases. Currently the nature of a bacterial infection is established by growing

29

the infecting bacteria until there are enough of them to identify with a microscope or with chemical tests. The time it takes to grow this culture can lead to a delay in effective treatment, with serious consequences. Chemists are now devising much more sensitive methods that will permit the identification in a few hours or less, so the appropriate medicine can be used. Sensitive methods are also being developed for the determination of the amounts of important hormones present in the blood, or of drugs used in treatment. The new techniques are revolutionizing the field of clinical chemistry.

Synthetic chemistry has a special structural role to play in medicine—producing materials that can be used anatomically. The chemistry of bone is understood, so new chemical materials will be made (some already exist) that can replace bones or teeth and that will be accepted by the body. Also, materials that can temporarily replace skin are needed for accident victims. Materials for replacement of veins or arteries, or even of body organs, need to be compatible with human biology, so they will not cause allergic reactions. Biocompatibility, which is also very important in contact lenses for instance (Figure 2.4), is an area in which chemistry will make increasing contributions in the future. The materials need to have the physical properties to fulfill their function, but they also need special surface coatings that look "biological" to the body.

In the continuing effort to understand the chemistry of life, exciting progress is being made in understanding how the brain works. It has been known for some time that nerve signals—in the brain and elsewhere in the body—are sent by

FIGURE 2.4

A contact lens. Modern synthetic materials make such medical devices possible.

special chemicals that pass from one nerve to another. In fact, this knowledge has been used by medicinal chemists to invent new pharmaceuticals for the treatment of various central nervous system diseases, such as depression or psychosis. Now progress is being made in understanding the nature of memory.

It is already clear that memory involves the biological synthesis of particular protein molecules, as well as physical changes in the brain cells, but there is still much to learn. In the future we can expect to understand as much about the chemistry of brain function, including short-term and long-term memory, as we currently understand about how nerves transmit signals. With this knowledge we may be able to improve human intelligence, certainly an exciting prospect.

There is a field called prebiotic chemistry that is concerned with how life could have arisen on earth, if indeed it arose by the spontaneous development of primitive chemistry into living chemistry. Understanding how this process might have happened could also help us understand what conditions would be needed for the development of life on other planets.

Some problems in this field have already been solved. For example, it has been shown that important chemical building blocks of current living systems—such as amino acids, sugars, and components of RNA and DNA—can be formed spontaneously from chemicals likely to have been present on the primitive earth. What is needed to promote chemical reactions is the ultraviolet light from the sun, electric discharges from lightning, and other conditions that must have been present in early times on this planet.

A particular problem in this field has recently been solved. Currently living creatures use enzymes to achieve the biochemical synthesis of RNA, and RNA to direct the synthesis of enzymes, which are proteins. How could such a cycle get started spontaneously, if each component is needed to make the other? Thomas Cech and Sydney Altman found—and received the Nobel Prize in Chemistry in 1989 for their discovery—that some RNA molecules can catalyze reactions just as enzymes do, although not as well. This finding suggests that the earliest life could have involved RNA acting both as the information molecule—the role it now plays—and also as the catalyst to turn that information into the production of new molecules.

According to this idea, there was an RNA world of life that came first. It was replaced by the current world,

31

in which RNA, DNA, and proteins now all play the roles that RNA alone once did. The new system is more effective, but the early one may have been enough to get things started. Fitting with this idea is evidence that the components of RNA can be formed spontaneously under the likely conditions found on the primitive earth.

Of course living creatures carry out their chemistry in cells. Chemists still need to learn how to imitate the properties of living cells with artificial systems, and also to learn how cells might have been formed spontaneously.

We can expect continued progress in developing systems that imitate the chemistry needed to start life. Such systems will not prove that life started that way on earth, but they may show that it is possible, on earth and elsewhere in the universe.

These are some of the topics in which knowledgeable chemists expect advances in the future. They are exciting and still challenging. Young potential scientists should consider joining the effort to achieve these and other important goals in the chemistry of health and life. There is still plenty of room for new ideas and new chemists interested in this important field.

More About Chemistry as the Useful Science

> No, a thousand times no; there does not exist a category of science to which one can give the name applied science. There are science and the applications of science, bound together as the fruit to the tree which bears it.
>
> —Louis Pasteur, 1871

In terms of the quotation by Louis Pasteur, the French chemist who invented pasteurization and immunization, the fruits from the tree of chemistry are abundant indeed. It is essentially impossible to point to all the products of chemistry that enrich our lives. It is not an exaggeration to say that the ways modern life differs most from that of earlier times resulted from our learning how to take the materials of nature and alter them to better serve us.

In this chapter, we will examine some of the roles chemistry plays in our everyday experience. Starting in the morning, we awake in houses and apartments constructed using chemical products. The furniture is constructed partly using modern materials, made possible by the chemical

FIGURE 3.1

Inside a plant where a useful chemical is manufactured. Careful design includes serious attention to the environment.

34

industry. We use soap and toothpaste designed by chemists, and we wear clothing made of synthetic fibers and synthetic dyes. Even the natural fibers, such as wool or cotton, are chemically treated and colored to improve their properties.

Our food is packaged and refrigerated for protection, and grown using fertilizers, herbicides, and pesticides. Farm animals are protected using veterinary medicines. Vitamins may be added to the food or taken as pills. Even the natural foods we buy, such as milk, must undergo chemical tests for purity.

Our means of transportation—cars, trains, airplanes—lean heavily on the products of the chemical process industries. The morning newspaper is printed on paper manufactured by a chemical process, with inks made by chemists, and illustrated with photography using film made by chemists. All the metal products in our lives come from chemically based refining and conversion of ores to metals, and of metals to alloys. Chemical paints protect them.

Cosmetics are made and tested by chemists. Weapons for law enforcement and national defense are based on chemistry. In fact, it is difficult to find any product we use in everyday life that is *not* based on chemistry and produced with the aid of chemists or their products (Figure 3.1).

What role does chemistry play in housing and home furnishings ?

Only wood, sand, and stone are natural construction materials, and they are held together and protected by synthetic chemical materials (Figure 3.2). Cement is a chemical product, as are adhesives such as those used in plywood and the metals used in nails. Glass is produced by chemists, and improved in products such as Pyrex to make it tougher. Paint is designed and created by chemists, and so are many modern solid materials. Plastics are synthetic, used in kitchen and bathroom appliances, in Formica and related materials, and in drinking glasses, plates, and utensils (Figure 3.3). Ceramics are made by chemists and used in kitchen and bathroom sinks and other fixtures. Metals are made from their ores by chemistry. Aluminum metal was

FIGURE 3.2

Products of the chemical process industries play an overwhelming role in home construction and home furnishings.

once a laboratory curiosity, but with a modern electro-chemical process it is now readily made from its oxide.

Carpets and draperies use synthetic fibers, at least in part, and synthetic dyes to color them (Figure 3.4). Refrigerators and air conditioners use special chemicals as cooling agents; gas furnaces and stoves can use either synthetic gas or natural gas that still undergoes chemical processing. Our homes are heated by gas or by oil produced by the petroleum industry, which refines and chemically modifies crude oil from nature. We insulate our buildings with synthetic chemical products, and use plaster or wallboard, siding and shingles, and floor tiles and carpets made of materials from the chemical process industries. The furnaces themselves, and the pipes that distribute heat, are made of chemical products—the metals, the insulators, and the ceramics.

Electricity comes into our homes carried by copper wires with insulation, both products of the chemical process industries. The electric outlets use plastic and metal, and they power incandescent and fluorescent light bulbs made entirely of chemical products.

Even the water that comes into the house is chemically purified to remove contaminants and disease germs. Before modern sanitation, involving chemical testing and purification, contaminated water was a major cause of human disease.

36

FIGURE 3.4

Dyes and fabrics are produced by synthetic chemistry.

What role does chemistry play in clothing ?

Shoe leather is made by chemically treating animal hides, and synthetic plastic materials often replace it. The rubber soles of shoes can come from rubber produced from the sap of rubber trees, but increasingly rubber substitutes such as the synthetic analog Neoprene are used. They are more durable and useful.

Synthetic fibers and synthetic dyes are in our clothing, and synthetic detergents and dry-cleaning fluids are used to maintain them. Of course there are also natural fibers, such as wool, flax, and cotton, that have advantages. However, even they are usually chemically treated, and always chemically dyed.

What about chemistry and transportation ?

Everything in an automobile is a product of the chemical process industries. The metals and paints are obvious, but the amount of plastic in a modern car is also very large. Special plastics are selected because they are strong and light in weight. Lighter vehicles consume less fuel.

The rubber in tires is altered by a process called vulcanization, in which the rubber molecules are chemically linked together to make the material tougher and more useful. The "rubber" hoses in the engine compartment are normally not rubber at all, but synthetic rubberlike materials that are much more resistant to oil and heat. The antifreeze is a synthetic chemical; the battery is a product of chemistry (*see* Chapter 4 for more about batteries); the upholstery is normally synthetic, or perhaps made of leather produced by chemical tanning, covering synthetic foamed plastics as seat fillings. The windows are safety glass, with plastic layers to prevent shattering. The fuels and lubricants are petroleum chemicals, with chemical additives to produce better antiknock behavior for the fuels, and better all-weather performance for the lubricants.

Sometimes useful fuels or lubricants can be obtained from crude petroleum by distilling it, but there are also chemical processes that change the natural molecules by the use of catalysts, "cracking" the large molecules in petroleum into the smaller ones for gasoline. Some of the large towers visible in petroleum refineries are used for distilling; others are used for those chemical processes by which more useful petrochemicals are produced. Lubricants have special chemicals added to them to give them better antifriction properties and to let them work well over a range of temperatures.

A modern application of chemistry is the use of catalytic converters in vehicle exhaust systems to reduce pollution. These use platinum, rhodium, and other substances to convert nitrogen oxides, carbon monoxide, and unburned hydrocarbons into less harmful chemicals.

Chemists are heavily involved in research and development in the automobile industry. In fact, at the research laboratory of one of the Big Three auto makers in the United States, chemists are the largest group of scientists. They are working to make better catalytic converters, and to reduce pollution by getting better burning of fuel. (Fire is itself a chemical process, whose nature is now understood.) Chemists are trying to change the painting process for vehicles so no organic solvents are used, they are trying to replace more of the metal in cars with modern plastics and ceramics, and they are working to improve the batteries so that electric cars might become more attractive. (This envi-

39

FIGURE 3.5

Essentially everything in an airplane, or in an automobile, is made from products of the chemical process industries.

ronmental aspect of automobile industry chemistry is discussed further in Chapter 4.)

Airplanes have special requirements (Figure 3.5). They would not be possible without lightweight strong aluminum. Special plastics and special fuels are used. Spaceflight requires even more specialized chemicals, including synthetic rocket fuels that produce very powerful thrust and special protective clothing made from synthetic materials. The next time you ride in a car or plane, or perhaps in a space shuttle, try to find anything that is not the product of modern chemistry. Unless you find a piece of wood or perhaps some cotton or wool you won't succeed, and even they are chemically treated and coated.

How does chemistry improve our food supply?

Some people who willingly use synthetic chemicals for many other purposes are unhappy at the idea that there might be synthetic chemicals in their food. This is completely understandable: no one wants to eat an untested new substance that might turn out to be dangerous. However, this proper concern can be taken to an extreme.

For example, we hear suggestions that people not eat anything whose chemical name they cannot pronounce. No one can survive such a rule. Glyceryl tristearate is the simplest chemical name for one of the components of fat found in natural foods. Glucosylfructose is not even the full chemical name for common sugar, and the chemical names of the vitamins—that we all need to eat in order to survive—are totally unpronounceable to someone who has not studied a little chemistry.

I propose a different idea: people who want to understand the chemical nature of their diets should study some chemistry. Then they will not only be able to pronounce complex chemical names but will also understand which chemicals belong in a healthful diet.

The role of chemistry starts right at the beginning of our food supplies, in agriculture. Chemical fertilizers supply the nitrogen and other chemical elements that plants need to grow. Without such fertilizers we could not produce enough food for our population. Weed killers are used to see that the food crops flourish, and insecticides are used to protect the crops.

The use of insecticides is one of the most controversial chemical activities. Killing the harmful insects is all very well, but will the insecticides kill birds, animals, and beneficial insects? Will they be carried on the food, and will they be harmful to humans when they are eaten? This topic will be discussed further in Chapter 4. Here it can simply be said that much effort now goes into research to develop better insect-control methods. Furthermore, chemists develop sensitive methods to detect pollutants in foods. The concerns are important, and they are being addressed.

The most toxic chemicals known are produced in nature, not in chemists' laboratories. Microorganisms such as bacteria make some very dangerous chemicals, and bacterial contamination of food is much more of a threat than agricultural chemicals ever were. Thus another contribution of chemistry is developing safe food preservatives and sanitary packaging to prevent bacterial contamination. Refrigeration, based on synthetic refrigerator chemicals, is another defense against food spoilage and bacterial contamination.

Chemistry also protects our water supply. Impure water can carry dangerous disease organisms, and in underdeveloped areas of the world there are often epidemics caused by unpurified water. Chemists have developed water treatment methods, using chlorine or ozone, that have probably contributed as much to human health as has the invention of modern medicines.

Medicinal chemists develop veterinary medicines to protect farm animals against disease. Chemists also develop testing methods to determine that products like milk meet our standards for purity and safety, and nutritional value.

Many of the foods we consume are not in their natural form, but are transformed. Grains are turned into breakfast cereals, cakes and cookies, and breads with the help of baking powder, a synthetic chemical product. The manufacturing processes are followed commercially with tests devised by food chemists. The fermentations by which beer, whiskey, and wine are prepared are followed with chemical tests.

We cannot remove "chemicals" from our foods—every substance in the world is made up of chemicals. We just have to be sure that the chemicals we use, whether natural or synthetic, are safe and effective for their purpose: protecting and enhancing our food supply.

What about national defense ?

Hardly anyone thinks that war is a worthwhile human activity, but if a war occurs everyone wants to win it. Thus governments have always called on scientists to make more effective weapons or better defenses. Chemistry has played a major role in both weapons and defenses.

The invention of gunpowder by ancient Chinese chemists completely changed warfare; modern ammunition uses newer versions of such explosives. They are chemicals or chemical mixtures that can undergo a chemical reaction with the release of a large amount of energy in the form of heat. The products of the reaction are gases such as nitrogen or carbon dioxide. The rapid expansion of the hot gases is the chemical explanation of an explosion.

The exception to this description is the explosion of a nuclear weapon. Again the explosion itself is the result of rapid heating of gases, but in this case the energy comes from a nuclear process in which the nature of the atoms changes, not from a chemical reaction in which atoms simply rearrange the ways in which they are attached. In a nuclear weapon the hot gases are just those of the air. The radiation that accompanies the explosion is also the result of nuclear, not chemical, processes.

Firing lead or steel bullets from steel guns—the metals produced by metallurgical chemistry, the bullets powered by chemical explosives—can certainly be thought of as "chemical warfare", but it is not the way that phrase is normally used. Instead, chemical warfare refers to the use of chemical poisons.

Bacteria use chemical warfare to kill other organisms, such as us, and very effectively in some cases. Humans are not nearly as effective. Poison gases have not played a large role in recent wars for a very good reason: they can kill both friends and enemies. Bacteria make poisons that kill other species, not themselves, but human folly is such that we want to kill other human beings who have the same biochemistry we have. Also, chemists have devised good defenses against poison gas, developing gas masks that react with and neutralize the poison, and protective plastic clothing for soldiers.

Modern chemistry plays a large role in the military, and every officer graduating from West Point must study chemistry. For this reason all branches of the U.S. military have

FIGURE 3.6

Composite materials, related to the "graphite" used in some tennis rackets and golf clubs, are used in the B-2 military airplane. Advanced technology, depending heavily on modern chemistry, is the chief military advantage of the United States.

programs to support chemical research in areas relevant to their needs. In the United States, the Office of Naval Research was actually the first government agency to set up such programs for support of important university research activities. The principal military advantage of the United States is our advanced technology, including that based on chemistry (Figure 3.6).

What about law enforcement?

Unfortunately, some aspects of synthetic chemistry have led to problems. Primitive chemical factories buried in the jungles are used to isolate raw opium from plants and convert it into heroin. Cocaine is extracted from plants, and its hydrochloride is chemically converted into crack, the free base of cocaine. Some other synthetic chemicals have also become drugs of abuse.

Legitimate chemists are attacking this problem in two ways. In one, sensitive new tests are devised to detect illegal drugs in the environment and in drug users. In another approach, medicinal chemists are developing new compounds that can block the action of addicting drugs; a good example is methadone, used to treat heroin addiction.

Chemical tests are playing an increasing role in other areas of law enforcement as well, developed by chemists working in the field called forensic science. Perhaps most striking is the identification of DNA as a way to prove that a sample of tissue or bodily fluid discovered at a crime scene actually came from the accused criminal. This chemical test requires modern methods by which DNA is copied hundreds or thousands of times using special enzyme catalysts, so a very small sample found in a crime scene can be amplified and identified. Then other enzymes are used to cut the DNA at specific places, giving a pattern that is characteristic for an individual. The evidence can be used to free an innocent person, or to indicate the likely guilt of another.

Remarkable modern plastics are used to make compact, lightweight bulletproof vests for police. Some chemical weapons are also helpful. Instead of shooting a criminal in order to apprehend him, police now have the choice of using powerful disabling chemical sprays, related to tear gas. Tear gas chemicals can also be used to flush a criminal from hiding. Correctly used, these weapons do not inflict permanent harm and are clearly better than a possibly lethal bullet.

Some crimes are committed by the mentally disturbed. Medicinal chemists have devised tranquilizers and other drugs useful in treating mental illness. These may make it possible for otherwise violent or irrational people to take up useful roles in society.

What does chemistry contribute to the quality of life?

"Man does not live by bread alone." With improvements in agriculture we do not have to spend every waking hour looking for food and shelter, but can add richness and variety to our lives.

Books, including this one, are printed with ink devised by chemists, and on paper produced by chemical processes. The cellulose in wood is surrounded by dark materials called lignin that must be removed to make paper. Special

FIGURE 3.7

The skates, and the athletic clothing, are made of mostly synthetic materials.

chemical procedures dissolve the lignin, leaving clean white cellulose behind.

Musical instruments are made using modern materials from chemistry. Even ancient instruments such as the violins produced by Stradivarius owe their exceptional musical character to the special varnishes devised by early chemists. Records, tapes, and compact discs are made of modern plastics, and television screens give off light because special chemicals glow, and in different colors, when they are struck by beams of electrons.

Photographic film is a plastic coated with chemicals that can be sensitized by light so they will undergo particular reactions when exposed later to developing chemicals. The chemistry of color photography is very elegant, as is that involved in instant Polaroid photos. These forms of photography were developed by building on quite advanced basic chemistry.

Swimming pools, football helmets, running shoes, in-line skates, skis, boats, fishing rods, and golf clubs all use modern plastics, modern ceramics and alloys, and modern paints and coatings (Figure 3.7). Even when we occasionally get back to nature, to get away from civilization and enjoy a simpler life, it is good to have the survival equipment and camping supplies made possible by modern chemistry and the products of the chemical process industries.

Considering the enormous range of topics in this chapter, it is impossible to cover the likely developments in all the fields. However, some advances will have a wide impact.

Modern Structural Materials

Many of the uses of chemistry involve novel solid materials, with improved properties. This is particularly true of what are called composites. These are combinations of materials in which the individual components cooperate. An example is the "graphite" used in some modern tennis rackets, golf clubs, and skis. It consists of bundles of graphite fibers (graphite is the black material in pencils) produced by high heating of a synthetic polymer, imbedded in a synthetic material such as epoxy resin. The resin helps hold the fibers together and protects them from breakage, while the fibers give strength to the composite that the resin alone does not have.

Composites are very old. In early civilization people learned that they could make bricks using straw as the fiber and clay as the other component, and that such bricks were not as fragile as those made from clay alone. More recently, fiberglass was invented using glass fibers held together with a synthetic resin, a polyester. The very thin glass fibers are flexible, unlike thick glass.

Modern composites are good, but not yet ideal. They are very strong and yet light in weight, so they are extensively used in aircraft and in automobiles. However, they do not stand up to high heat, and some are hard to machine. It is

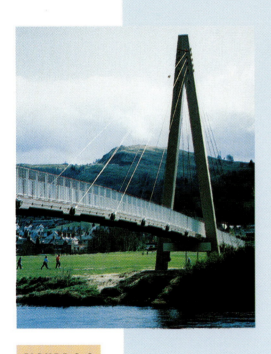

FIGURE 3.8

A pedestrian bridge in Scotland constructed using fiberglass composites instead of metal.

46

clear that the future will bring composite materials with greatly improved properties and lower cost. We can expect to see the metal in vehicles increasingly replaced by such materials. Furthermore, they may increasingly find use as structural materials in bridges, for instance, in which strong lightweight materials are needed (Figure 3.8).

The advances in composites are guided not merely by random experimentation. Modern techniques let chemists understand the detailed molecular structures inside a composite and how to optimize them.

Ceramics, such as one sees in cups and saucers, have many desirable properties, but they can be fragile. Composites can be made by imbedding graphite or polymer fibers in the ceramic, and they are then much less breakable. Ceramic composites can even be used for very high temperature applications, such as in rocket engines for spacecraft.

For many materials the exposed surfaces are particularly important. Thus a material with useful bulk properties—either a simple solid material or a composite—can be made with a surface coating of another material to give the special surface properties needed. Early examples of surface coating include paint and the chrome plating of steel. Modern materials chemists are developing other special materials for surface coatings. Since the surface layer is often quite thin, an expensive material can be used if its properties are worth it. An interesting example is the development of methods to apply a thin coating of diamond onto metals, which then become very tough and scratch-resistant. This can be thought of as a special example of an important general goal—indestructible paint that protects metals parts for many decades.

The field of chemistry is increasingly concerned with the properties of organized mixtures of substances, not just of pure substances such as nylon or Bakelite. This shift in focus can be seen as an orderly intellectual transition. At first chemists looked at complex materials, such as the wood in trees, and separated them into their components. They identified the chemical nature of cellulose, and of the lignin in wood that is removed when wood is converted to paper. Now there is increasing interest in the chemistry of the interacting components, not just the pure substances. This new emphasis is also true in biochemistry, where chemists first had to identify all the molecules of life and their properties. Now there is increasing interest in the chemical interactions among these molecules within the cell.

47

Many devices already have complex arrangements of different chemicals. Photographic film at one point consisted simply of silver chloride crystals imbedded in a synthetic plastic, but modern color photographic film has many layers containing different dye chemicals. An integrated-circuit chip, which makes modern computing possible, consists of many layers of specialized materials on a pure silicon wafer. If the chemistry can be mastered, even better devices are envisioned for the computers of the future. In addition, there is one other area in which advances will revolutionize our use of electricity.

Superconductors

Normal electrical conductors, such as copper wire, resist the flow of electricity to some extent, turning part of the electric power into heat. It has been known for many years that a few such conductors become **superconductors** at very low temperatures. In a superconductor there is no electrical resistance, and thus electricity is transmitted with complete efficiency. It would clearly be better if our power lines were constructed using superconductors, so no electricity is wasted.

In addition, there are potential applications of superconductors in magnets and in electronic devices. In fact, some scientific magnets are already constructed with present-day superconductors. They must be cooled close to the lowest possible temperature—called absolute zero (−273 °C, −459 °F)—using liquid helium as the coolant, before they become superconducting.

This need for extremely low temperatures is a problem. It is simply not practical to cool transcontinental power lines with liquid helium. What is needed is a superconductor that operates at higher temperatures. The ideal is a room-temperature superconductor, but even one that operated at the temperature of liquid nitrogen (−196 °C) would be useful. The air is 80% nitrogen, and turning it liquid by compressing it is not very expensive.

Recent breakthroughs in materials chemistry suggest that practical superconductors operating at such temperatures may be within reach. Interestingly, among the best so far are special ceramics, materials like porcelain but containing very specific extra chemicals. The remarkable thing is that these become superconductors at easily accessible temperatures, but above those temperatures they are actually insulators. They are not even electrical conductors except when they are cooled into the superconducting state.

If these materials can be improved, and this is an area of very active chemical research, they could usher in an era in which power could be generated where convenient, and transferred throughout the country very efficiently. Another area in which we can expect great advances relates to our food.

Food Supplies

Chemistry plays a role in the production of food, with modern agricultural methods and animal health products, and in the processing of food for our consumption. Chemistry also brings us more knowledge about what is in our foods, and what should be in a healthful diet.

With improved agricultural chemicals we should be able to increase the productivity of farmland (Figure 3.9). Insect control will be achieved using selective environmentally friendly new insecticides, some based on the detailed chemistry that insects themselves use. (We will explore this topic further in Chapter 4.) Fertilizers will better meet the needs of food plants, and weed control will help ensure that vital sunlight, fertilizer, and water go only to the food crops (Figure 3.10).

We are now much more sensitive to the matter of contamination of food. Chemists will continue to test

FIGURE 3.9

Chemistry plays a vital role in food production, with agricultural chemicals such as fertilizers and other products used for food protection and storage.

FIGURE 3.10

An airplane spraying a herbicide on a cotton field, to control weeds.

50

food and help defend it against spoilage and bacterial contamination. Improved packaging makes it possible to store many foods without refrigeration, and this technology will expand further in the future. Intelligent concerns about the effects of intentional and unintentional food additives will also be addressed.

New agricultural chemicals will be developed to replace those that raise questions. For instance, farm animals are frequently treated with antibiotics to decrease disease, but we don't want those antibiotics to end up in our foods. Some otherwise effective antibiotics are now banned for use in food animals because of this concern, and they will be replaced with improved medicinal agents.

Food spoilage can be prevented with antioxidant chemicals, but those that should be used are the ones that also bring human health benefits, as some do. Food colorings increase the appeal of some foods, but not if they have questionable effects on the consumers.

We need to develop new agricultural and food chemicals that have no unwanted side effects. Now that this requirement is completely clear, there is very active research to develop the products of the future. Only recently have we all become sensitive to these matters. They represent a tremendous challenge and opportunity for modern chemistry, and for the chemists of the future.

Chemistry and the Environment

Hurt not the earth, neither the sea, nor the trees.

—*The Bible: Book of Revelation*

Our environment is under pressure from an increasing human population and the products of an advanced civilization (Figure 4.1). As a central part of that advance, chemistry certainly contributes to the problem; it also holds the keys to solving it. In this chapter we will address a basic question.

Can we have a modern society, based on advanced technology, that does not hopelessly pollute the earth?

The answer is yes, but only if we pay attention to some limits. For one, we can afford to populate the earth, but we cannot afford to overpopulate it. This is a question of human behavior, but even here chemists can help.

Medicinal chemists have been active in developing new birth control agents, but the ideal ones are not yet in hand. We need methods that are cheap, convenient, safe, and effective, and that meet the religious and cultural requirements of our various populations. At one time this was an area of active research, but it is sufficiently controversial that few pharmaceutical companies are currently pursuing it.

Ironically, the population problem has been caused to some extent by chemistry. The human death rate was once so high that overpopulation was no threat. Great disease scourges swept the world, such as the plague or influenza epidemics. Medicinal chemistry has greatly increased our survival rate, so now we have a population problem. Surely we prefer controlling our population by choice, not by massive deaths.

The second limit on our activities is that we must take full account of their environmental effects. We must not assume that the oceans or the air or the soil can simply absorb whatever waste products we add to them.

FIGURE 4.1

As human activities and technical advances improve our life span and the quality of our lives, we must be careful that they do not damage the environment on our quite special planet.

Environmental chemistry is one of the fast-growing areas of science. It is concerned with understanding what happens to various materials when they are added to soil, water, or air. With increased understanding have come better procedures and products. For example, plastics are now being made—for use in packaging, for instance—that readily degrade in the environment. As another important approach, chemists are learning how to recycle plastics, metals, and rubber so the considerable investment of energy and materials that went into their manufacture is not lost, and so they don't pose a disposal problem. New chemicals have been invented for use in refrigeration and in air conditioning that do not pose a threat to the ozone layer that protects us from harmful high-energy ultraviolet sunlight.

We must package food, to keep it sanitary and fresh, but with environmentally safe materials. We need refrigeration for our food supply, and air conditioning to open up hot regions to settlement, but not with a refrigerating chemical that exposes us to cancer-causing radiation from the sun. We also need to manufacture products without causing environmental damage.

53

Can industry manufacture chemical products cleanly ?

The chemical process industry is the largest segment of manufacturing in the United States and in some other advanced countries. Materials are converted to useful products by performing chemical changes, but waste products are usually also produced. What is done with them holds the key to clean manufacturing.

At one time waste products were simply dumped into rivers, buried in the ground, or vented into the air. We assumed, or hoped, or pretended to believe that this would cause no problem. Sometimes we call them "chemical waste products", but since every substance consists of chemicals this is not a real distinction. Now the United States is involved in a massive cleanup of toxic waste sites, places where nature has not simply taken care of the problem. We now know we must solve the problem ourselves.

Chemical manufacturers who belong to the Chemical Manufacturers Association, and who make 90% of the

FIGURE 4.2

The Responsible Care symbol of the Chemical Manufacturers Association.

chemicals in the United States, have adopted a program called **Responsible Care**, in which they pledge to manufacture without causing environmental damage (Figure 4.2). The results are often amazing.

As one example, there is a chemical plant in Tennessee that uses its Responsible Care program to guide its operations from an environmental perspective. It constructed the nation's most innovative wastewater treatment facility that is elevated 7 feet above the ground to make protecting the groundwater easier. Some of the ash from its coal-fired boilers would have been buried in former times but is now used to make concrete blocks, a good pollution prevention success. It also has the nation's only commercial facility to convert coal into chemicals eventually used in such items as photographic film and synthetic fibers. By using locally available high-sulfur coal, it conserves about 1.5 million barrels of oil annually. In addition, the facility recovers 99.7 percent of the sulfur contained in the coal. That chemicals-from-coal facility is so innovative, it was even designated a National Historic Chemical Landmark in 1995.

54

FIGURE 4.3

A plant of the Eastman Chemical Company boasts the only commercial chemicals-from-coal facility that converts coal into chemicals used in everything from cars to fabrics. The process conserves about 1.5 million barrels of oil annually.

Another pollution problem has to do with solvents, liquids that are used to dissolve various chemicals. When organic solvents evaporate, they pollute the air. Increasingly manufacturers are learning how to carry out their processes using water as the solvent. Water can easily be cleaned up and returned to the environment, as in the Tennessee chemical plant just described. Water-based paints, including those used on automobiles, are replacing paints containing organic solvents.

These environmental concerns reflect the feelings of the manufacturers—who live with their families and raise their children in the same world as the rest of us—but they are also driven by environmental regulations. Careless manufacturing is cheaper. Thus reasonable environmental laws are endorsed by the leaders of responsible companies, so they are not at an economic disadvantage with respect to irresponsible competitors.

What can chemists contribute to solving the problem of radioactive waste from nuclear plants?

Not all waste products can be safely turned into useful products. Chemical toxicity can be neutralized with more chemistry, but only time can make radioactive materials safe. With current technology, there seems no way to avoid producing radioactive waste in nuclear energy plants. And yet, it seems that nuclear energy is clearly in our future. In some countries it is already a major energy source. What can be done about the waste?

Actually the physical amount of radioactive material produced in a nuclear plant is not large, but it is imbedded in a lot of other material. Currently the whole mass is put into storage, but the amount is so large that this causes difficulties. Chemists are working to solve the problem.

We generally know how to separate one chemical from another, by taking advantage of their different properties. Chemists are now trying to learn how to isolate and concentrate only the radioactive material. It will have to be stored for long times, but the problem is much smaller if the harmless other material does not have to be stored as well. We may eventually even learn how to do something useful with radioactive waste materials once they are isolated.

What about pollution from routine human activities, such as driving a car?

When gasoline is burned, some poisonous carbon monoxide is given off and some of the nitrogen of the air is converted to nitrogen oxides. Unburned gasoline products are also emitted. These can be offensive in themselves, but in sunlight they combine to form irritating chemicals that are part of smog. Every major city has some problem with smog, and for a few cities it poses real health threats. What can chemists do to help?

The first step was simply to learn what the chemistry was that led to smog. With this information, countermeasures were invented. The chemistry of combustion was investigated, and engines were modified to reduce the amount of pollutants they produce. To handle the rest, catalytic converters were installed in the engine exhaust. These solid catalysts convert the pollutants to less harmful materials. However, this is not the final answer.

56

A prototype electric car. Better batteries are needed to make such cars generally practical.

An electric vehicle, powered by an electric battery, is really nonpolluting. Furthermore, electricity can be generated—for instance, using water power—without pollution. Why do we not yet have major fleets of electric cars and buses (Figure 4.4)?

The problem is the batteries. A storage battery is a device that stores electricity chemically. As a chemical change occurs in the battery, electricity is produced. When the battery is exhausted, because all the chemicals that produce the electricity are used up, it can be recharged. Electricity is passed into the battery again, reversing the chemical change and regenerating the chemical substances so they can again produce electricity. It is easy to describe the battery needed to make an electric vehicle practical, but it doesn't yet exist.

We need a battery that can hold enough electricity to let a car go 300 miles or so before a recharge is needed. Then we need to be able to recharge the battery in 15 minutes or less, at a "gas station", so we can be on our way. The battery has to last for a year or more before it is replaced, depending on its cost. Furthermore, the battery must not be so heavy that it adds a lot to the weight of the car. The requirements are not impossible, just difficult. So far they have not been met.

Developing batteries for electric cars is a very active area of research. Materials are known that can carry the amount of electricity we need in a reasonable weight and size, but so far they don't work well in storage batteries. This is still a problem for the future.

Another possible approach involves what is called a fuel cell. Batteries that cannot be recharged, such as those commonly used in flashlights, consume some material to generate electricity; normally that material is a metal such as zinc or lithium, which gives up electricity as it is converted to a new chemical compound. It is also possible to generate electricity in reactions that consume other materials as fuel.

For example, fuel cells are known that generate electricity from the reaction of a hydrocarbon, such as gasoline, with air. The process can be thought of as cold burning: the same products are formed as in combustion, but at the low temperature of a hydrocarbon fuel cell there are few pollutants formed.

Even more interesting is a fuel cell operated with air and hydrogen. The product is simply water, whose vapor can be returned to the air. The problem here is with the hydrogen.

Hydrogen gas is very dangerous; with air it can form explosive mixtures. At one time it was used in balloons, including dirigibles, but the danger meant that it was replaced with helium. (Helium is chemically inert; it won't react with oxygen under any conditions.) For safety, hydrogen will have to be stored in some form other than as a compressed gas if we are to use it in fuel cells for automobiles. For example, some metals such as nickel can absorb large amounts of hydrogen and release it when needed.

The problem has not yet been fully solved, but when we solve it we may move to what has sometimes been called the hydrogen economy. Hydrogen will be generated from water with electricity at an electric generating plant, and then transported in a safe form (still to be discovered). We will use it in cars to generate electricity and power electric motors, while the hydrogen is converted back to water. The net result is to transport electricity from an efficient generating plant into our vehicles, carried in the form of hydrogen. With further chemical research, we will get there.

Can we control insect pests without harming valuable creatures?

We need to control certain insects, those that bring disease or destroy our food. Termites and carpenter ants can even destroy our houses. However, without bees many plants could not reproduce. How can we control the harmful insects selectively, without poisoning ourselves or other harmless creatures?

DDT raised many of these questions to public awareness. It is a very effective synthetic insecticide, and its use to control mosquitoes led to a major decrease of malaria in tropical countries, where it is a serious health problem. The effect was so important that the inventor of DDT, Paul Müller, received the Nobel Prize in Physiology and Medicine in 1948 for this work. And yet, DDT is now used very little. It is banned in the United States and many other countries.

DDT is selectively toxic to insects, and does not kill other creatures. However, it does accumulate in some birds, and leads to weak eggshells so the embryos do not survive. This side effect is clearly not acceptable, and it explains the DDT ban. With the decrease in use of DDT worldwide, there has been an increase in malaria in tropical countries.

Is this our only choice: Kill the birds, or accept a devastating disease for humans? Chemists working in this area say no. They are developing selective pesticides that will control harmful insects without causing undesirable side effects. Even more promising is finding a way to take advantage of the special chemistry that insects perform.

Insects use chemicals to signal each other for mating or to lead to food supplies. A female insect will give off a chemical signal to attract males of the same species, and other insects are not attracted. That is, each species has its own chemical code. Chemists are now learning what the code is for various species, and using it for insect control. For example, traps can be put out carrying the mosquito chemical code, to attract and capture male mosquitoes. It is sometimes enough just to put out the chemical code itself, without the trap, so the males will be too confused to find a female for mating.

As we learn about the ways in which the chemistry and biochemistry of insects differs from that of other creatures, and individual insect species differ among themselves, we can expect success in our efforts to control insects that pose a threat to us. We don't want to upset the balance of nature, but neither do we want to accept a world in which some insects cause human suffering or disease. With modern approaches we should be able to reach an appropriate compromise.

59

How can we be sure that new useful chemicals do not harm the environment, or have undesirable biological effects?

Because pesticides and weed killers are designed to have useful biological effects, they are carefully scrutinized to be sure that they do not do unintended harm. This is also true for new medicines, which must pass rigorous tests for possible undesired side effects. However, only in the past few years have chemists, and others, become sensitive to some of the surprising undesirable side effects of other new chemicals.

CFCs and the Ozone Layer

Perhaps the most striking example of an undesirable side effect was connected with refrigerators and air conditioners. These devices need to contain a gas that can be compressed into its liquid form. This compression causes the material to expel heat—the heat that refrigerators give off outside the cooling compartment and that air conditioners give off to the outside air. When the liquid is allowed to expand into a gas again, it absorbs heat into the cooling coils that are inside the refrigerator, or on the cooling side of an air conditioner. What gas should be used?

At one time refrigerators operated using ammonia or sulfur dioxide, both of them dangerous gases if inhaled in large quantities. There were some deaths when those early refrigerators leaked. Thus chemists set out to invent a new chemical that would have the right properties. It had to absorb a lot of heat when it expanded from the liquid to the gaseous form, but it had to have no harmful biological effects. After a lot of research, the CFCs—chlorofluorocarbons—were invented for this purpose, and they were seemingly ideal.

They really were harmless biologically, to such an extent that they became widely used for other purposes as well. For instance, their property of expanding from a liquid to a gas when the pressure is released made them ideal for producing foamed plastics as insulating materials, and they were used in spray cans for paint, shaving cream, and even food products such as whipped cream.

The first hint of trouble came from some fundamental chemical studies on the molecule ozone, a gas made up of three linked oxygen atoms. F. Sherwood Rowland and Mario Molina—who shared the Nobel Prize in Chemistry in 1995 with Paul Crutzen for these discoveries—found that chlorine atoms were very powerful catalysts for the decomposition of ozone in the upper atmosphere. In this decomposition, an ozone (O_3) molecule and a free oxygen atom (O) are converted to two molecules of ordinary oxygen (O_2).

As Figure 4.5 shows, the chlorine atoms catalyze this decomposition by taking the third oxygen atom from ozone, producing one molecule of O_2 and a reactive species, chlorine oxide (Cl—O). This Cl—O then transfers its oxygen atom to a free oxygen atom to make another molecule of O_2. After this transfer the chlorine atom is able to carry out the sequence again, so a few chlorine atoms can

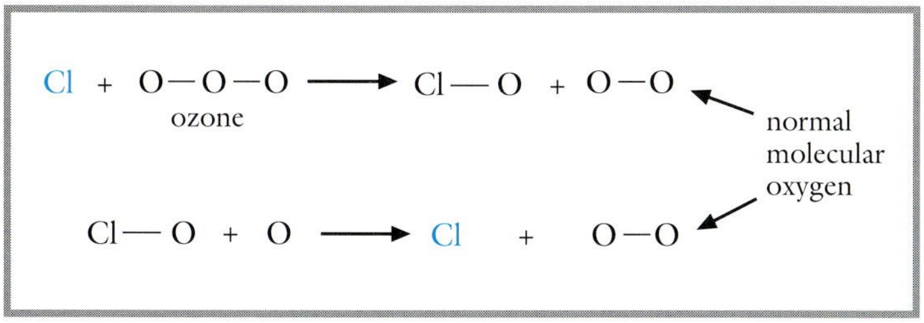

FIGURE 4.5

A chemical sequence of two reactions by which a chlorine atom, Cl, can convert an ozone molecule and a free oxygen atom into two molecules of ordinary oxygen. The Cl is regenerated, so the sequence repeats itself many times.

destroy a lot of ozone molecules—as many as one thousand. (Eventually the chlorine atom combines with other species to stop the sequence.) The effect is not only to destroy ozone molecules; each sequence of two reactions also prevents the formation of another ozone. The free oxygen atom, produced when high-energy ultraviolet radiation hits an O_2 molecule, would normally add to O_2 to make an O_3 molecule, but the Cl—O captures the oxygen atom and prevents its reaction with O_2.

Ozone down at ground level is actually harmful, contributing to smog and destroying many common materials. However, the ozone in the upper atmosphere is very important—it absorbs the particularly high-energy ultraviolet light from the sun that can be very dangerous to all life on earth, including human life. What does this have to do with CFCs?

It turns out that the same high-energy ultraviolet light can break up CFCs to release chlorine atoms. From what was learned about the rate of this process, chemists concluded that enough chlorine atoms were likely to be produced, from the amount of CFCs that we were sending into the atmosphere, to pose a danger to the ozone layer. At the same time, studies of the atmosphere revealed that the ozone layer was being temporarily destroyed near the South Pole of the earth, just where the CFCs in the atmosphere were concentrating (Figure 4.6). The problem occurred in seasons in which the ultraviolet light was particularly strong, and thus particularly likely to liberate chlorine atoms from the CFCs.

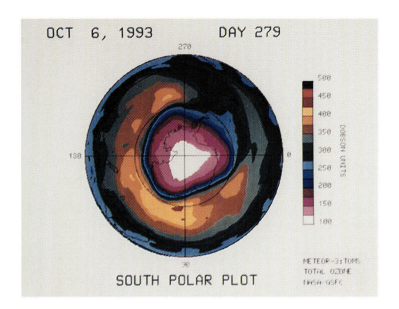

OCT 6, 1993   DAY 279

SOUTH POLAR PLOT

FIGURE 4.6

The hole in the ozone layer over the South Pole, shown in white. Based on data from a NASA satellite.

These facts led to the strong suspicion that CFCs, harmless as they might be biologically, were not acceptable because of the new risk that had not previously been thought of seriously. As a result, new chemicals have been invented for use in refrigerators and air conditioners, and others are being introduced for powering spray cans and manufacturing foamed plastics. These new chemicals do not have the same ability to destroy the ozone layer that CFCs have, but work is still under way to find the ideal substitutes.

The lessons from this saga are important. First of all, chemicals are not like people—new chemicals that will be introduced widely should be considered guilty until proven innocent. This is the way all new medicinal chemicals are treated, and it is the way other chemicals must be tested. We need to become as sophisticated as possible about all the complex ways in which chemicals can interact with our world and its inhabitants, and this needs serious research in the field of environmental chemistry.

The second lesson is that we can't simply abandon all progress because of uneducated fear. Would we be better off if refrigeration had never been invented? Would we be better off if we were still using the early poisonous refrigerating chemicals? The CFCs solved all the problems we knew about when they were invented. When new problems emerged, chemists addressed them by inventing new chemicals that did the same job that CFCs did, but without causing the problems. This is the way progress has been made in the past, and it is the way of the future. The critical need is to learn as much as we can, so we don't get too many surprises—undesired effects that we never thought of.

Not All Undesirable Chemical Effects Are Lethal

Another lesson we have learned is that some chemicals, even if they are not lethal poisons, can have effects on humans or other living things that are not desirable. A good example is DDT, whose undesirable effect was to interfere with the ability of some birds to reproduce. New medicinal chemicals are tested to see if they are lethal to humans, of course, but they are also screened for other nonlethal effects such as increasing the possibility of human birth defects. Any such side effect dooms an otherwise promising medicine.

There is now worry that some other chemicals that are in our environment may also have undesirable effects. In particular, there is concern that some chemicals may imitate female hormones and interfere with the reproductive system of men. At the time this book is being written, the scientific picture is cloudy. It is not clear whether there are indeed human changes in reproductive ability, or whether synthetic chemicals could be at fault. It is clear that large doses of some synthetic chemicals can have damaging effects on the reproductive systems of animals, but extensive studies are now under way to see if small amounts of such chemicals are a problem for humans. It will be several years before the full story is clear. However, even if it should turn out that the problem has been exaggerated, the concern should still be taken seriously. Any new chemical that is introduced into the environment, or into human contact, should be the subject of careful testing based on our developing understanding of what the possible negative effects are.

Just as medicinal chemists can make newer compounds that have no side effects, after testing reveals a problem with a proposed medicine, so too can industrial chemists make

63

new agricultural chemicals or new packaging materials, for example, to avoid any problems that testing reveals for their own potential products. Our increasing sophistication lets us learn more and more about what kinds of tests a new substance must pass before it is ready for general use.

In the past there was sometimes a tendency to assume that any new chemical that was not a poison, or did not cause cancer, would not be a problem. Modern science knows better, and modern chemistry can meet the requirements of our new knowledge.

THE FUTURE

The careless days are over. Responsible Care is the watchword for the future, not just in chemical manufacturing but in all human activities that might damage our world. However, no sensible person advocates returning to the lifestyles of cavemen, or even of our ancestors of previous centuries. Instead, we need to use what science has learned, and will continue to learn, to make continued contributions to human welfare. We need more effective medicines, safer and more abundant food, and all the other benefits that chemistry has brought us. However, we must also understand and solve any related problems.

This is an excellent time to be in the field of environmental chemistry. Scientists working in this area are in demand and have the wonderful opportunity of "doing well by doing good". They will be designing beneficial alternatives to replace products that are not environmentally friendly. They will be designing better methods to detect the presence of undesirable materials in air, water, soil, and food. They will be designing better methods to manufacture useful compounds. As Chapter 7 will describe, there will be increased emphasis on learning how to avoid the formation of harmful side products during chemical reactions.

One of the biggest challenges relates to what can be called limited stability. We need chemical products that do not fall apart during their normal use, but we do not want chemicals so stable that they persist forever in the environment. The outstanding chemical stability of the CFCs was considered to be one of their great advantages, but we now understand that this stability lets them rise

64

undestroyed into the upper atmosphere where they cause damage. Some of the substitutes for CFCs have as their main advantage that they are destroyed in the lower atmosphere before they get up into the ozone layer where they can cause problems. Building in just the right amount of environmental stability is a challenge that chemistry can meet.

A challenge that is just now being addressed concerns what has been called the **greenhouse effect.** Certain gases, such as carbon dioxide or methane, can act like the glass in a greenhouse—letting the sun's light through, but blocking the resulting heat from radiating back into space. This means that such gases, released into the atmosphere, could lead to global warming.

This story is more complex than was the relationship between CFCs and the ozone hole, and the detailed science is not as clear. Future work will establish whether there is indeed such a problem, and if so what to do about it. The difficulty is that carbon dioxide is a product of the burning of coal or petroleum or wood. If it is a real threat to our climate, we will have to introduce drastic changes in the ways we generate energy. It will be ironic if nuclear energy, that has engendered such fears, should turn out to be safer for our future than is the burning of fossil fuels. More scientific evidence is needed before we can evaluate the magnitude of the proposed greenhouse effect, the magnitude of the threat that it poses, and the magnitude and nature of the changes that are needed to deal with the threat.

In the next century, it seems clear that chemistry will make even greater contributions to our lives than it has in the past. However, it will be done with full understanding and full concern that our improvements come without damaging the natural world. As the quotation at the beginning of this chapter suggests, the alternative is unacceptable.

65

Computers and Chemistry, and the Electronics Revolution

The universe...stands continually open to our gaze, but it cannot be understood unless one first learns to comprehend the language and interpret the characters in which it is written. It is written in the language of mathematics....

—Galileo, *Il Saggiatore* (1623)

Chemistry has made numerous contributions to the electronics revolution, especially to computers. However, computers have also brought great new strength to chemistry. There is a major field called computational chemistry, and it is rapidly growing. The bulk of this chapter will be devoted to describing this field, and its prospects.

What has chemistry contributed to the modern electronics revolution that makes computers possible?

Specialized materials are critical to modern electronics. Indeed, they have played a role in even earlier developments in electricity.

For example, Thomas Edison maintained a private chemistry laboratory all his life, and he was able to invent a practical electric light bulb only after he tried many possible materials for the filament. His earliest success was a filament of cotton carrying carbon particles; modern incandescent light bulbs use tungsten. Tungsten, one of the chemical elements, combines useful chemical properties with important physical properties as a material. It conducts electricity, but with enough resistance to become very hot and give off light. Other specialized chemicals are part of the coatings or the gas in fluorescent lights.

Computers existed in the days of the vacuum tube, with its slow speed and high power needs, but they were awkward unwieldy objects. In the early 1950s such a computer generally filled a good-sized room. It had less computing power than the simple scientific calculators that students can now hold in their hands. The revolution came with the invention of the transistor. This device, like the vacuum tubes it replaced, acts to amplify small electrical currents and is a critical component of most electrical circuits.

Transistors take advantage of the special properties of the element silicon. It is a poor conductor of electricity, but

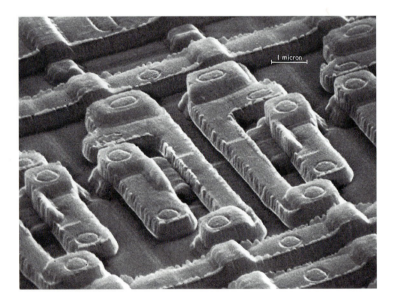

1 micron

FIGURE 5.1

A portion of a logic circuit on a silicon-based computer chip, magnified 23,000 times.

its electrical properties can be changed by adding small amounts of other elements. Some of these elements, such as phosphorus, can incorporate into the silicon and add extra electrons. Others such as boron incorporate and add a "hole" into which an electron can flow. Each change makes the silicon a much better electrical conductor, but with a difference. Electricity tends to flow away from the silicon containing added elements that give it extra electrons, but electricity flows into the silicon with a hole into which electrons can go. With the proper arrangement of such materials, one can construct a transistor to amplify electrical currents.

Many transistors and other electrical circuit elements (e.g., resistors and capacitors) can be incorporated on a single small silicon chip in an integrated circuit (Figure 5.1). Because these parts of a complex circuit are very close to each other, signals travel between them with great speed. The power requirements are also very low compared with circuits using vacuum tubes. The transistor, based on silicon, is the heart of our modern computer revolution. It is no accident that the region where many of the modern developments in electronics and computers take place is called Silicon Valley.

The electrical properties of other specialized chemical materials are also important. Conductors and insulators play a general role, and superconductors, which can conduct electricity without energy losses caused by resistance, are of current use and of great future potential. Chemists have made specialized magnetic materials for information storage in computing, and other materials for the construction of compact discs (CDs) for computing and other uses (Figure 5.2).

This area of chemistry is part of a field called materials science. Electrical engineers work in it,

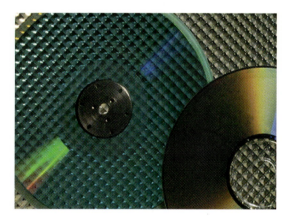

FIGURE 5.2

Modern chemical materials make compact discs possible, used both for music recording and for computer information storage.

69

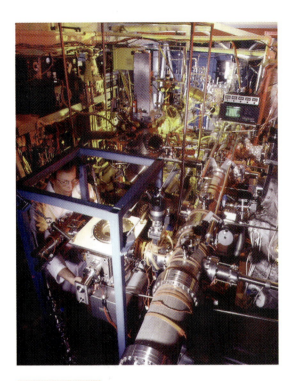

FIGURE 5.3

Research on semiconductors at the Research Triangle Institute in North Carolina. Not all chemical research is done in test tubes and flasks.

as do solid-state physicists and chemists. They are designing and creating the specialized materials needed for applications in electronics (Figure 5.3).

Almost as large as the impact of chemistry on computers has been the influence of computers on chemistry. All of the instruments that chemists use for measurements are computer-controlled, and the experimental results are generally fed into a computer to be analyzed mathematically. However, there is also a growing field in which the "experiments" themselves are done simply in a computer. The chemical structures are manipulated and the properties of unknown chemicals are predicted using computation alone. This is the field of **computational chemistry**

What is computational chemistry ?

In the early part of this century a remarkable new theory was formulated: **quantum mechanics**. According to this theory, all of chemistry can be calculated from first principles without the need for any more experiments. The properties of any unknown molecule can be accurately predicted, and the rate and products from any unknown chemical reaction can be predicted. All that is required is to solve a deceptively simple equation, the Schrödinger equation:

$$H\psi = E\psi$$

Although the equation looks simple enough, the definitions of H, E, and ψ require a level of sophistication in mathematics, chemistry, or physics far beyond what is assumed in this book. Unfortunately, so far this equation can be solved exactly only for the very simplest molecules.

Even so, a number of ways have been invented to obtain approximate solutions of this equation for cases of real interest. The best of these approximate methods need lots of computer time, but with faster computers and larger memories, and with new computer programs written by computational chemists, progress has been rapid. It is increasingly possible to reproduce experimental data by calculation, and to make useful predictions.

Another important computer technique is called molecular mechanics. This method of calculating the properties of molecules is particularly useful in molecular modeling, by which the best three-dimensional geometries of flexible molecules are predicted. It is heavily used in modern medicinal chemistry to design new drugs, and is also used to predict the geometries into which proteins will fold.

How does computational chemistry help us determine the shapes of proteins

Once the sequence of a gene has been determined, it tells us the sequence of amino acids in the protein that is coded for by the gene. For instance, if one of the three-letter words in the gene is GGG it codes for the simplest amino acid, glycine. However, the properties of proteins, such as enzymes, depend on their three-dimensional geometries. In an enzyme, catalytic groups that need to be close together to cooperate may be attached quite far apart in the sequence. For example, in the enzyme ribonuclease folding brings two important amino acid groups together that are separated by more than 100 other amino acids along the chain. Unless we can predict the three-dimensional folded geometries from the sequence, that sequence will not tell us enough (Figure 5.4). Molecular mechanics addresses this problem, with help from quantum mechanics. There are two different types of solutions that it can find.

The simplest solution is found if the folding of a protein leads to the most stable three-dimensional structure. Then all that is required is to figure out what that most stable structure is. That is a big challenge for a molecule that may contain 10,000 atoms, but it is easier to address than the other possibility. In the second general case the protein might adopt a stable geometry, but it might not be the best one, just the easiest arrangement to get to. A good analogy has to do with mountains and valleys.

If you were descending from a mountain, you might come to rest in a valley surrounded by other mountains, and decide that this valley is low enough and is where you

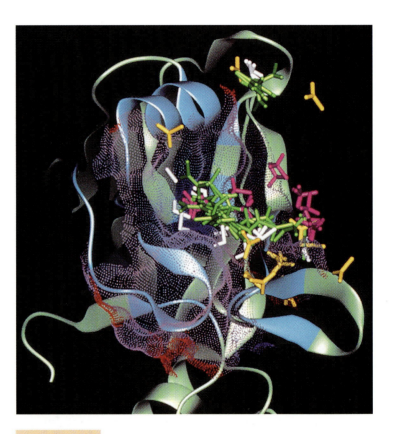

FIGURE 5.4

A protein is shown in the ribbon diagram that is commonly used to represent its general shape, without the chemical details. The ribbons indicate the ways in which the flexible chains of the protein fold up. Such images are currently based on the direct determination of the protein structure using X-ray methods, but there is increasing hope that computer methods can predict the final three-dimensional structures into which particular proteins fold.

will stay. However, over the next mountain range there could be a valley that is even lower, but because the first one was easy to get to, you decided not to press on. Molecules can do the same thing. They may settle into a stable geometry that has low energy, but not find a way to get to an even lower energy state. The calculated geometry of the unattained lowest energy state will then be irrelevant. To address this problem, computational chemists must calculate not only the energies of various stable states, but also the pathways that the protein is most likely to follow while folding up.

As if this were not problem enough, there is evidence that the folding of some proteins is assisted by what are called chaperone proteins. These bind to the unfolded protein, guide it along a desirable path (how like a human chaperone), and then release it at the end. Thus they are a type of catalyst for folding. In calculating the most likely folding path, any chaperone involved would have to be included in the calculation.

The challenges in computational chemistry are substantial, but great progress is being made. For young scientists with an interest in computers, computational chemistry is a field with an exciting future.

73

THE FUTURE

Electronic Materials

New materials with improved electrical, magnetic, or optical properties will continue to pace advances in electronics. There is great excitement about potential advances in the field of materials science, with cooperative research by chemists, physicists, and electrical engineers among others.

The chemistry involved in the silicon-based chip will be changed and improved with new chemistry to permit even more speed and miniaturization. New magnetic materials will be developed to play a role in computer memories, letting us store more information in smaller spaces. Chemicals that change color when exposed to one kind of light, and then change back again with another kind of light, have great interest as another way to store information in computers. They might be part of a different kind of "electronics", in which light is used instead of electricity in what could be called optical computers. Light moves much more rapidly than does electricity. Chemists are also working to construct improved lasers, the likely light sources for such novel computers.

A single molecule is a very tiny object, even if it has lots of atoms. Still, modern techniques permit chemists to "see" individual molecules in some cases, and even to determine the color of a single molecule. Is there a way to store information by changing a single molecule, using light or electricity? Is there a way to read that information? If so, and this is still just a dream, we could pack a huge amount of information into a small space.

The effort to construct smaller and smaller electronic devices and computers is part of a new field called nanochemistry. Chemists working in this area are trying to construct electronic, optical, and mechanical components that are truly miniature. For example, many bacteria have tiny "motors" that help them swim. Can we make motors of that size?

Another important related area is molecular electronics. Chemists hope to create useful electronic circuits on a truly miniature scale by having the components undergo what is called self assembly. In chemistry we do not generally force molecules to come together and link up; this normally happens spontaneously. Can we use such spontaneous linking to assemble electronic components in the desired ways? Can we produce useful polymers that conduct electricity as well as metals do, so such polymers can take the place of metals in electronic circuits? Can we assemble such circuits by using the ability of light to promote some chemical reactions, so that light beams focused to very small images can produce the desired circuits?

Focused light is already used in part in this way. Chemical reactions promoted by light are used to produce polymers that shield the silicon wafers and direct the chemical etching of these wafers to produce the circuits. The hope is that it may be possible to replace this somewhat indirect process with the direct production of circuit elements by photochemistry.

Liquid crystals are extensively used in displays on wrist watches and portable computers. They consist of molecules that orient in an electric field, and whose optical properties depend on this orientation so they can translate electrical signals into visible images. This is a field in which we can expect continued progress.

Another exciting future advance concerns molecules whose electrical or optical properties change when other chemicals bind to them. Such molecules could be used in chemical sensors, to detect toxic compounds or monitor the level of hormones and other biological compounds. They will in a sense be part of electronic noses or electronic taste buds, responding with a signal when the chemical binding event occurs.

Computational Chemistry

Many young people today want to pursue careers that will make use of their expertise and interest in computers. Computational chemistry is one of the best places to do this.

Experimental chemists increasingly understand that there are problems they cannot solve except by computation. We must develop better and faster ways to perform the very large calculations required. Along the way, computer methods will make predictions that experimental chemists will check. When it is found that the computer is right every time, we can then trust its predictions.

What are the questions that computational chemistry already addresses, with incomplete success?

75

1. The bonds that hold atoms together in a molecule are flexible, so a given molecule can have a variety of shapes (somewhat like a worm, twisting and coiling). What is the most stable shape of a given molecule? This question is of interest for large molecules such as proteins and nucleic acids, but even small molecules such as sugars need reliable calculations of their shapes if we are to understand their properties.

2. Even if we know the most stable shape of a molecule, the less stable shapes may also play a role in its chemistry or biology. When drugs bind to a protein, some of them change their shape in order to fit. How much energy does it take to make a molecule take up a different shape? How will this affect its ability to bind into the right spot, if it is a drug?

3. How will the shape of a molecule differ if it is dissolved in water, compared with its shape as a component of a pure chemical? This question relates to the influence of solvents, such as water, on chemistry. Since biological chemistry occurs in water, we need to understand such influences.

4. What chemical properties will an unknown molecule

have? If we are making it as a fuel, for instance, how much heat will it give off when it is burned? How rapidly will it undergo some unknown chemical reaction? In this area rapid progress is being made, and calculations are increasingly able to predict the results of experiments.

5. How does a chemical reaction occur? We want to see the atoms and molecules coming together, watch the speed at which chemical bonds are made and broken, and see how the energy is carried off by the products. Except for a few simple cases we don't yet know how to observe a chemical reaction directly—it goes on too fast, and on a tiny scale. However, it can all be calculated. The question is, are the calculations correct? In this area as well, the calculations are being checked by their ability to predict correctly such things as the speed of a reaction, and how that speed changes as the temperature changes. Progress is real, but the goal line is still some distance down the field.

6. When two chemicals react, what is the most likely product? For example, a chlorine atom can remove a hydrogen atom from most molecules that have one or more carbon–hydrogen bonds. Which hydrogen will be removed? Will the reaction be completely selective for that hydrogen, or will a mixture of products be formed?

These are some of the questions that computational chemists hope to solve. They are not simply of academic interest. If we can predict the properties of unknown chemicals, or the products that will be formed from unknown chemical reactions, we can save a lot of fruitless experimenting. The old "shake and bake" era of trial and error will be replaced by rational planning that we can trust. Young chemists with ability and interest in computing will play a major role in creating this new era.

CHAPTER 6

Catalysis in Living Organisms and in Industry

> The meeting of two personalities is like the contact of two chemical substances; if there is any reaction, both are transformed.
>
> —Carl Jung, *Modern Man in Search of a Soul*

Jung's assertion may be true of human interactions, but it is not true of chemical reactions when one of the chemical substances is a catalyst. A catalyst is able to induce a chemical reaction in some substance it contacts without itself being permanently transformed. This chapter will describe some important examples, and how they work.

No subject so pervades modern chemistry as that of catalysis. Industrial manufacturing of useful chemical and medicinal products uses catalysts to perform the needed chemical reactions. The biochemical reactions that constitute life, in chemical terms, are catalyzed by enzymes that are among the most effective catalysts known. One of the major functions of DNA and RNA is to produce the code that indicates which enzyme the cell shall make, and translating that code into action involves the action of specific enzyme catalysts. Catalytic converters in automobiles help destroy nitrogen oxides and carbon monoxide in the engine exhaust.

What do catalysts do?

In general terms, catalysts make chemical changes go faster. Also, the catalyst does this without undergoing permanent change itself, so it can act over and over again. A tiny amount of a catalyst can process an enormous amount of material, until the catalyst is ruined in some way. (Chemists say that the catalyst has then been "poisoned". Real poisons usually work by ruining enzymes, the body's catalysts.) Many chemical reactions are slow because they must pass through intermediate states that are of high energy and that are thus hard to get to. Afterward the products are of low energy again, as were the reacting molecules.

Once again, a good analogy has to do with mountains, this time containing tunnels. Going from Denver to San Francisco by land should be a downhill coast, from 5000 feet down to sea level, but in between there are mountain ranges that require energy to surmount. They slow down progress. We have solved this geographic problem by digging tunnels through the mountains so there are new pathways that do not require all that energy and effort. Catalysts create new pathways for chemical reactions, avoiding the high-energy states that are difficult to pass over. In a sense, they create tunnels. Chemists now understand how they do this.

A REAL EXAMPLE

Ethyl alcohol (the alcohol in beer, wine, and spirits) can be converted to diethyl ether (which is used as an anesthetic in surgery) in a reaction catalyzed by an acid such as sulfuric acid. Chemists write the process in the following way:

$$R-OH + R-OH \rightarrow R-O-R + H_2O$$

Here the **R** stands for the ethyl group, whose chemical formula we don't need to write out here. The O is an oxygen atom and the H is a hydrogen atom. The reaction equation says that two molecules of ethyl alcohol, **R**—OH, combine to form a molecule of diethyl ether, **R—O—R**, plus a molecule of water. However, even though this reaction is a favorable process, it does not occur at a significant rate without a catalyst. There is no danger that a bottle of vodka would start to contain ether on standing.

The sulfuric acid catalyst can add an H atom to an O atom (actually it adds a proton, which is an H atom without its electron and is written H^+). It is easy to see how this might help. To form water (H_2O) one of the oxygen atoms of the ethyl alcohols needs to pick up a second hydrogen atom, and the sulfuric acid catalyst adds it by an easier path than if it had to come from the other alcohol molecule. In order to form the ether molecule, the second **R—OH** group must lose its hydrogen atom. Thus a catalytic hydrogen atom is formed again, although it is not the same hydrogen that was originally added.

Let me show this reaction scheme the way chemists would really write it, with the details (see Figure 6.1). Chemical studies show that this scheme is correct. In this reaction one carbon–oxygen bond has to break to form the product water, a difficult process needing lots of energy. However, it is much easier to break that bond if the oxygen carries an extra H^+, a proton supplied by an acid catalyst. The two ethyl alcohol molecules link up in what is called a displacement reaction: a water molecule is displaced by an oxygen of the second ethyl alcohol. Then a proton is lost to form the product diethyl ether, and the catalytic proton is ready to go on to another ethyl alcohol and repeat the process.

The energy mountain that the uncatalyzed reaction would have to cross is so high that the reaction does not occur at an appreciable rate. When the catalyst is added a new pathway is created (the tunnel through the mountain). Catalysts always work by introducing new paths for otherwise more difficult reactions.

What about enzymes, the catalysts of life ?

Living cells such as ours do not contain strong acid catalysts. Still, enzymes use schemes very similar to the one just outlined, but with some important special characteristics.

Enzymes are proteins, made up of hundreds of linked amino acids. When they carry out a catalyzed reaction they first bind the reacting molecule, called the substrate, into a cavity in the surface of the enzyme. Some enzymes then add a proton to the oxygen atom of a substrate, just as in the

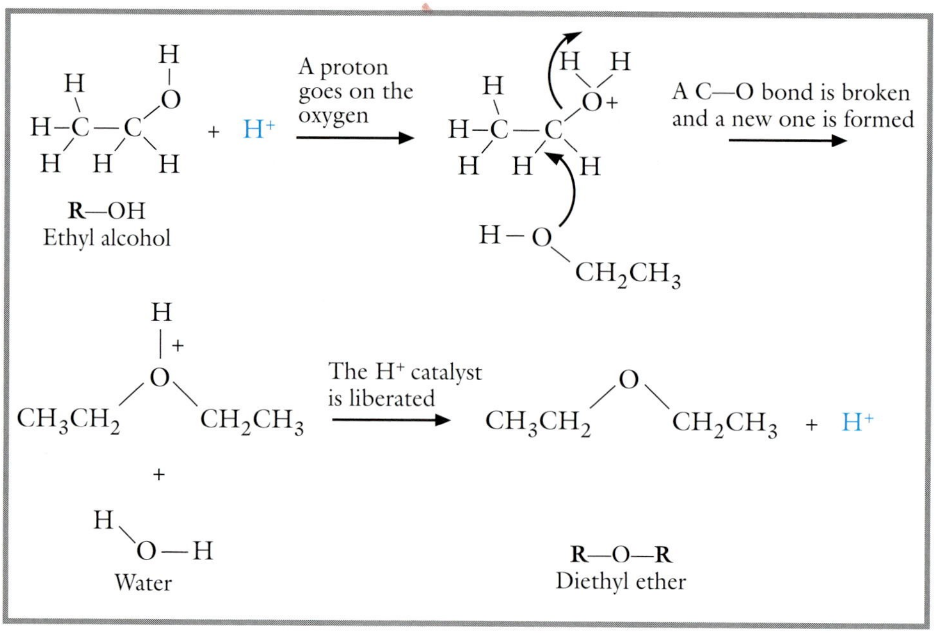

H
|
H   O
 \ |
H–C—C   + H⁺   →   A proton goes on the oxygen

H H H

R—OH
Ethyl alcohol

(H H)
   \ /
    O+
H |
 \ |
H–C—C   A C—O bond is broken and a new one is formed →

H H H

H—O
      \
       CH₂CH₃

H
| +
O
CH₃CH₂   CH₂CH₃   →   The H⁺ catalyst is liberated   →   O
                                                              CH₃CH₂   CH₂CH₃ + H⁺

+

H
 \
  O—H
Water

R—O—R
Diethyl ether

FIGURE 6.1

The conversion of ethyl alcohol to diethyl ether catalyzed by a proton from an acid. There are two kinds of arrows here. The straight ones going from left to right just show how the reaction proceeds, from starting materials through intermediates to products. However, the little curved arrows are commonly used by chemists to show which bonds are being made and broken in some step of the sequence. In the second step, a water molecule is breaking away from a carbon atom of an ethyl group (the top curved arrow) while a second molecule of ethyl alcohol is forming a new bond to that same carbon atom (the bottom curved arrow). The water molecule breaks away carrying off the two electrons of the C—O bond, so it becomes a neutral water molecule without a positive charge. However, the attacking oxygen atom of the second ethyl alcohol molecule uses some of its electrons to form the new bond, so the oxygen atom becomes positively charged. Only after it loses the H⁺ in the last step is neutral diethyl ether formed. In this drawing the ethyl group is shown in two different ways. The first one shows each carbon atom with its attached hydrogens, to make it clear that the oxygen is attached to carbon and that the second ethyl alcohol molecule is reacting at that carbon. However, when this detail is not needed the ethyl group is drawn as CH_3CH_2 or CH_2CH_3.

conversion of ethyl alcohol to diethyl ether shown in Figure 6.1. This proton would come from a chemical group in the enzyme, derived from one of the amino acids that make it up. Later another proton would be taken back to regenerate the enzyme, so it could continue to catalyze reactions when the product breaks away and a new substrate binds.

The details of enzyme catalysis are increasingly understood, but not as well as the details for simple chemical processes. With the technique called X-ray crystallography, chemists can get a detailed picture of the chemical structure of an enzyme, sometimes with a substrate bound to it. Chemists can also change the chemical structure of the substrate and of the enzyme. Such a change can be done either with chemical reactions or by changing the gene that codes for the enzyme, a process called mutagenesis. The effect of these changes on the reaction rates can help us understand what is going on. Of course the ideal would be to watch the reaction occur, or take fast movies or video films of it. This is not yet possible, so indirect methods, such as those just discussed, have to be used. In Chapter 8 we discuss more about how chemists learn the details of chemical and biochemical reactions.

How fast are biological reactions catalyzed by enzymes, and how selective ?

It is not at all unusual for an enzyme to speed up a chemical reaction by a factor of 10,000,000,000. If you took 5 seconds to read the last sentence, you would have taken 10 billion times as long, 1500 years, without the acceleration of your enzyme-catalyzed reactions. This is a huge effect. It makes life possible. Another special feature of enzyme catalysis is its selectivity.

A simple chemical catalyst such as sulfuric acid can speed up lots of different chemical reactions, but this lack of selectivity is normally not a problem for chemists. We can add the catalyst to a reaction mixture containing only the chemicals we want for the desired reaction. However, enzymes must operate in living systems with hundreds of potential reacting chemicals in the cell. Enzymes have to be selective to carry out only the needed reactions.

The selectivity results from the fact that enzymes bind their substrates before catalyzing a reaction. The enzyme pocket into which the substrate binds has a well-defined shape that will accept some substrates but not others. It even has handedness (as discussed in Chapter 1), which permits it to bind a natural amino acid but not its mirror image. Furthermore, the groups in the enzyme that help to bind the substrate have defined positions, so even substrates that might fit into the pocket will not bind there if they cannot interact correctly with the binding groups.

Finally, the groups in the enzyme that actually catalyze the reaction also have well-defined positions. If the wrong substrate binds, the catalytic groups of the enzyme are not able to reach the important atoms.

This selectivity helps the enzyme catalyze the reaction of only the desirable substrate, but there is another form of selectivity as well. A given substrate may have several possible reactions that it could undergo; the enzyme will select one that leads to the desirable product.

Some enzymes catalyze the selective breaking apart of peptide bonds in other proteins, the bonds that link amino acids in a protein. This is often needed to produce another enzyme. For example, the enzymes that digest the meat we eat are produced in an inactive form and are turned into active digestive enzymes when a few of their own particular peptide bonds are cleaved. The cleaving enzymes can select these bonds for cutting, but not others. If the wrong peptide bonds were cleaved, the digestive enzyme would be destroyed. The precise geometry of substrate binding into the pocket of the cleaving enzyme holds only one peptide bond in the right position to be cleaved.

The handedness of a product can also be directed by an enzyme. For example, natural amino acids are put together from simple molecules that are not different from their mirror images—so they have no handedness—but the enzymatic reactions make only one of the two mirror image possibilities (*see* Chapter 7). This selectivity again reflects the geometry of the enzyme–substrate complex.

Do vitamins play a role in biological catalysis?

Not every chemical reaction can be catalyzed well only by the simple chemical structures found in proteins. Some enzymes use additional catalysts to help, called coenzymes. These are typically derived from vitamins.

For example, vitamin B_1 (also called thiamin) is vital for our conversion of sugars to energy and important biological chemicals, but humans do not make thiamin in our cells. Instead, we obtain it either from our foods or as vitamin supplements. Chemists have learned how to manufacture thiamin that is identical in every respect with the vitamin obtained from nature; it is used in vitamin pills and in

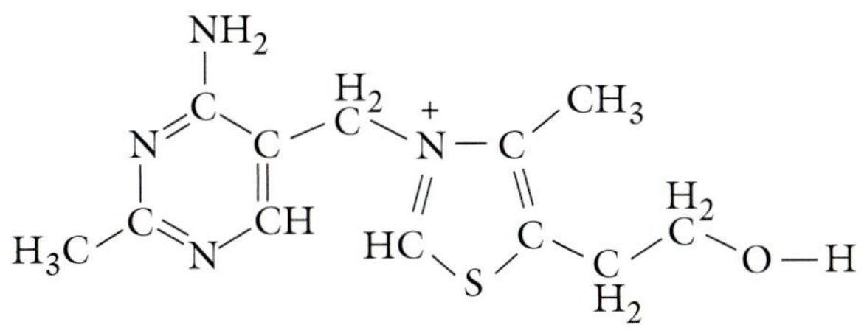

FIGURE 6.2

The chemical structure of thiamin, which is also called vitamin B$_1$. In living systems two phosphate groups are attached to the O–H group at the right. The result is a coenzyme, called thiamin pyrophosphate, that helps many important enzymes catalyze biochemical reactions.

enriched foods.

The thiamin we eat is converted into a coenzyme in our bodies by adding two phosphate groups, catalyzed by an enzyme specific for that purpose (Figure 6.2). The extra phosphate groups help the thiamin bind into a pocket of another enzyme—an enzyme that requires thiamin pyrophosphate to help with the catalysis. Then the substrate of the reaction binds next to the thiamin pyrophosphate, and the enzyme helps the thiamin derivative catalyze a reaction of the substrate. The products are released from the enzyme, and it is ready, along with its thiamin-derived coenzyme, to carry out another reaction.

All the B vitamins function in this way, by being converted to coenzymes that help in enzymatic catalysis. For example, the ethyl alcohol in beer or wine, produced by fermentation, results from the action of an enzyme using modified thiamin (vitamin B$_1$), followed by a different enzyme using modified niacin (vitamin B$_3$).

How much do we understand about how enzymes work?

The detailed three-dimensional structures of many enzymes have been determined by X-ray methods, often with a substrate bound into the pocket so we can see exactly where it goes (Figure 6.3). Also, the catalytic groups have

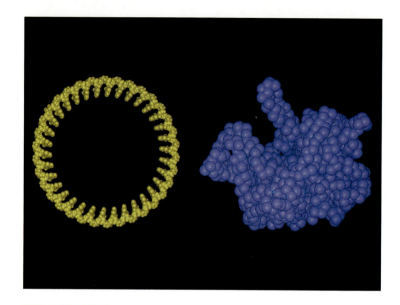

FIGURE 6.3

A DNA circle above an enzyme that acts on it. The structure of the enzyme was determined by X-ray crystallography.

been identified in most cases. However, there are still gaps in our understanding.

We cannot yet account for enzyme catalysis in quantitative terms. That is, we cannot yet specify how important the various characteristics of an enzyme are to its ability to speed reactions by 10,000,000,000-fold or so. In addition, we cannot yet construct artificial enzymes, using the same chemical groups as in natural enzymes, that achieve the same catalytic effectiveness. Chemical research is addressing these problems. Indeed, there is a field called biomimetic chemistry—a term coined some years ago by the author—that focuses on imitating natural enzyme chemistry and achieving both better catalysts and better understanding. Some interesting details are now understood. One has to do with the flexibility of enzymes.

When we described the binding of a substrate into the pocket of an enzyme, we implied that the pocket already had a defined shape that the substrate had to fit. This is not quite correct. It has been demonstrated for many enzymes—and is believed to be true for all—that the pocket originally has a rather open shape, letting the substrate get into it easily, and that the enzyme pocket then closes around the substrate to make a tight fit.

Originally people talked about a substrate fitting an enzyme the way a key fits a lock, but now we know that the Venus flytrap is a better model. This plant folds up around an unfortunate insect. The folding is triggered by the presence of the fly, just as the folding of the enzyme is induced by the presence of the substrate. After the catalyzed reaction is complete, the enzyme opens up again to let the products leave and to prepare for the next substrate.

Another detail has to do with the ability of enzymes to carry out several steps rolled into one. When a substrate is bound to an enzyme, there are several enzyme groups held nearby that can help carry out the catalytic reaction. It is now clear, from evidence using deuterium (the heavy isotope of hydrogen), that they can operate simultaneously. Reactions in which deuterium is moving are usually slower than the same reactions involving normal light hydrogen. From the details of the rate effect with deuterium it is often possible to say that two deuteriums are moving at the same time.

For example, in the reaction converting ethyl alcohol to diethyl ether, with which we started this chapter, we showed two steps in which protons were being moved. In the first step, we put a proton on the oxygen of one ethyl alcohol molecule. Later, after the displacement reaction, we removed a proton from the oxygen of the other ethyl alcohol molecule. If an enzyme were carrying out this reaction—no enzyme is known that can do this particular reaction—it would put the proton on with one catalytic group, hold the two ethyl alcohol molecules together by binding both into a pocket, and remove the other proton with another catalytic group, all at once.

There is a big rate advantage in such simultaneous processes; it seems that the ability to perform such all-at-once reactions is one of the secrets of enzyme catalysis. Chemists have made some biomimetic catalysts that imitate this feature, and they are quite effective.

Is our knowledge about enzymes useful in designing new medicines?

Many drugs act by blocking specific enzymes. The drug is designed to look like the normal substrate, but with a difference. The drug does not have a normally reactive chemical group like that of the substrate, so the enzyme cannot catalyze any reaction of the drug. Furthermore, the drug is designed to bind more strongly than a normal substrate can. This keeps it sitting on the enzyme, preventing access by normal substrates. This will slow or completely block undesirable biochemical reactions.

There is a rational way to design a strong-binding analog of a substrate: take advantage of what is known about how chemical reactions occur. As I described previously, ordinary chemical reactions are slow because they have to proceed through intermediate chemical structures of very high energy (the top of the mountain). A catalyst lowers that energy by producing a new path (the tunnel) in which the intermediate structures are stabilized by binding strongly to the catalyst. Thus one way to make a strong-binding analog of a normal substrate is to make the medicine look like one of the intermediate high-energy structures involved in the reaction. With this trick new drugs have been made that bind thousands of times more strongly than the substrate does.

How do solid metal catalysts work, such as those in automobiles?

The fundamental principle of catalysis operates with these catalysts also: introducing a new path in which the ordinarily unstable intermediates are stabilized. This stabilization occurs because of the special properties of certain metals.

Consider catalytic hydrogenation, by which hydrogen molecules add to unsaturated organic molecules—which contain some carbons linked by double bonds to other carbons, shown as two lines between them—to saturate them (Figure 6.4). In "partially hydrogenated" peanut butter, for example, carbon–carbon double bonds in some of the fats are removed by adding hydrogen. The same process is used to convert unsaturated oils to margarine, and to remove the unsaturation in soap so it will not turn rancid. The reaction occurs because it is favorable, with the products lower in energy than the starting materials, but the rate is essentially zero in the absence of a catalyst.

The problem is that hydrogen gas has two hydrogen atoms strongly bonded to each other. That bond is not pre-

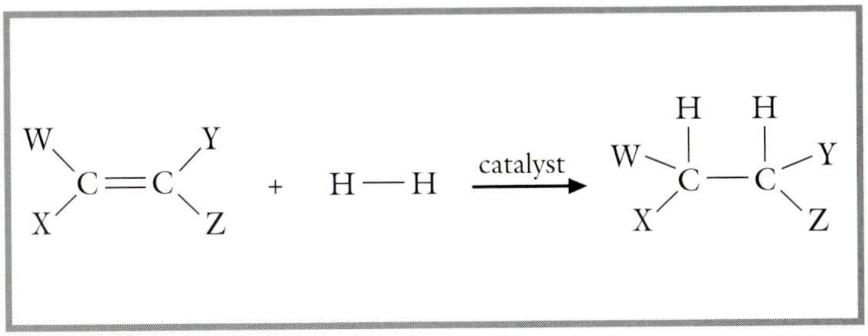

FIGURE 6.4

Hydrogenation of a carbon–carbon double bond. The groups W, X, Y, and Z can have various structures.

sent in the product, and the two hydrogens are added individually to different carbon atoms. However, it is difficult to break the strong hydrogen–hydrogen bond so the reaction can occur; the catalyst solves the problem.

Hydrogen molecules react with metallic nickel, forming new nickel-to-hydrogen bonds in which the two hydrogen atoms are no longer attached to each other. They can then add separately to the two ends of a double bond. Thus metallic nickel is a catalyst for hydrogenation, and the one commonly used.

As we mentioned in Chapter 1, the periodic table organizes elements according to their common properties. Similar elements lie in a column above and below each other in the table. Thus it is no surprise that catalytic hydrogenation can also be catalyzed by metallic palladium and metallic platinum. These elements lie below nickel in the periodic table.

The special ability of metals like platinum to form bonds to hydrogen atoms and other chemical groups makes them useful catalysts for various reactions. Platinum finds major uses in industrial reactions, such as those involved in the petroleum industry. Other metals are also used as catalysts for industrial chemistry, since the properties of different metals let them catalyze different chemical processes. For example, in the catalytic converters used in automobiles to reduce pollution from engine exhaust the special properties of metallic rhodium make it particularly effective in removing nitrogen oxides. It is a problem of current interest to find another catalyst that is as effective as rhodium, but more available and cheaper.

Why does industry often use solid catalysts that don't have the special properties of metals ?

In chemical manufacturing, such as the conversion of petroleum to useful chemicals, there is sometimes a need for an acid catalyst. It serves the same function that we described in our first example, the conversion of ethyl alcohol to diethyl ether. That is, the catalyst adds a proton to a reactant to permit an easier path to form products. However, there is an advantage to using a solid acid catalyst that does not dissolve in the reacting chemicals—one does not have to remove the catalyst from the products. Typically, a solid acid catalyst is packed into a tube and the reacting chemicals pass through the tube, either as liquids or as gases. The tube is often heated. The products come out the end of the tube and the solid catalyst stays in place. This is what goes on in some of the large towers you see in oil refineries.

There is another special advantage to some solid catalysts— they may have cavities that give shape selectivity to the reactions they catalyze. This is true of a group of catalysts called zeolites, compounds made up of the elements aluminum, silicon, and oxygen. Chemists have made various zeolites that have

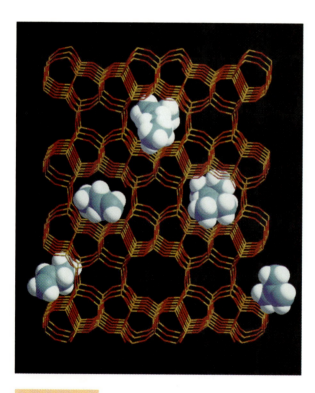

88

FIGURE 6.5

A computer model of a zeolite catalyst with spaces into which a hydrocarbon molecule can fit. The shape of the spaces promotes the formation of a four-carbon molecule from petroleum.

spaces inside their solid structures able to hold some molecules, not others. For example, some zeolites will hold a slim molecule but not one with bulges that make it too fat to fit the space in the zeolite (Figure 6.5).

These catalysts select for reacting molecules, and products, that can be accommodated in the spaces. In that sense, they are much like the enzymes discussed earlier, although their chemical composition is completely different. Some of this shape selectivity is used in the petrochemical industry, to guide reactions along desired paths (Figure 6.6).

Part of a petrochemical plant, in which petroleum is converted to useful chemicals. The towers contain catalysts that break crude oil down to smaller molecules that can be used to make plastics, antifreeze, and other useful materials.

 T H E   F U T U R E

Catalysis will be even more important in the future. All biochemical reactions occur with the help of natural catalysts, the enzymes. Enzymes speed reactions enormously, permitting chemical reactions to occur under very mild conditions (at body temperature and in neutral water solution) that would otherwise require strong heating and extreme reaction conditions. Furthermore, enzyme catalysts are very selective. The desired compound is produced in high chemical yield, without the formation of undesirable waste products. These important qualities of enzymatic reactions lead to two branches for future research in chemical catalysis.

In one branch, we will increasingly use enzymes themselves to carry out desirable chemical reactions.

$$N_2 + 3\,H_2 \longrightarrow 2\,NH_3$$

FIGURE 6.7

The conversion of nitrogen from the air into ammonia, a critical chemical for making fertilizers and many other important chemicals. The catalysts used commercially to carry out this reaction require high heat and pressure, but microorganisms on the roots of living plants can convert nitrogen to ammonia at room temperature and atmospheric pressure. One of the future challenges is to develop commercial catalysts that can also operate under these conditions, saving energy.

They can often catalyze reactions different from those normally involved in their biochemical functions, and they can be modified to expand their capabilities. For instance, chemical reactions on the enzymes themselves can introduce new catalytic groups. Another choice is to do genetic engineering, changing the gene that codes for the enzyme so that a modified enzyme is produced.

The second branch of the future uses enzymes as role models. Enzymes tell us what is possible, and increasingly work on the detailed chemistry that enzymes perform will tell us how they do it. Then chemists will design new molecules that can do the same thing but that, unlike enzymes, are not based on proteins at all.

For example, enzymes are selective for the shapes of the molecules that they bind to, but chemists have made the mineral materials called zeolites that are also able to bind molecules and select for the correct shape. Enzymes use relatively simple catalytic groups, those that are found in amino acids and in a few coenzymes, but chemists have found that very effective catalysts can be based on some metals, such as platinum, that are not used at all in enzymes.

There is a particular biomimetic challenge that could have very important results for humankind. The nitrogen gas of the air is quite unreactive, but highly available. If it can be induced to react with hydrogen it forms ammonia, NH_3, which is used to make fertilizers and many other chemicals that contain nitrogen (Figure 6.7). Currently this reaction between nitrogen gas and hydrogen gas is performed at high temperatures and very high pressures using some metal catalysts. And yet, some living plants can take nitrogen from the air and convert it to ammonia at room temperature and atmospheric pressure. This biological conversion, known as nitrogen fixation, is achieved with enzymes that are better catalysts than the catalysts we use in industry so far. Can we learn how the enzymes do it, and more important, can we learn how to do the same thing? This challenge still waits to be solved by future chemical research.

Another challenge has to do with imitating photosynthesis. Can we find catalysts that will let us use sunlight to split water into oxygen and hydrogen? Can we catalyze the use of sunlight to make other molecules? Can we take carbon dioxide from the air and catalyze its conversion into useful chemicals, as plants do in photosynthesis?

There are other challenges that are unrelated to imitating natural catalysis. We need to learn how to catalyze the conversion of methane, CH_4, into larger molecules such as are present in gasoline. We need to develop catalysts that will direct the formation of polymers with high selectivity for the desired properties. We need to develop chemicals that can catalyze their own formation; such self-replicating molecules will imitate the ability of biological organisms to reproduce.

Increasingly we can expect that new catalysts will be designed by chemists to achieve enzyme-like speeds and selectivities. Increasingly such new catalysts will make it possible to perform chemical synthesis and manufacturing with high efficiency. Costs will drop, energy consumption will drop, and environmental problems will decrease. This is one of the most exciting areas for future advances in chemistry. It will lead to increased understanding of biochemistry, increased understanding of chemistry, and an ability to greatly improve the processes we currently use to manufacture important chemicals.

91

C H A P T E R 7

How Chemists Create New Molecules

Concern for man himself and his fate must always form the chief interest of all technical endeavors...in order that the creations of our mind shall be a blessing and not a curse to mankind.

—Albert Einstein, 1931

Modern chemists are amazingly prolific in the design and creation of new chemical substances. Most of them are made only in small quantities in laboratories, so their properties can be evaluated. In this chapter we will discuss why chemists make new molecules, and how they do so.

Before we get to this topic, consider how different chemistry is in this respect from most other sciences. Where is synthetic astronomy, in which new stars or planets are created so their properties can be compared with the natural ones? Where is synthetic geology, in which different versions of the earth are made to see whether they are better or worse than the one we have? Biology has always had a synthetic component, with plant and animal breeding to create new food plants and new strains of livestock. Still, chemistry remains the leader by far in its concern with all that is possible in the world, not just that which is found in nature.

How do chemists decide what new molecules to make ?

We must distinguish between the purposes of the work and the choice of molecules to serve those purposes. One of the purposes is to make new molecules that may be useful. In line with the quotation from Albert Einstein at the beginning of this chapter, sometimes chemists make new molecules because they might be effective medicines or because they might serve the wide variety of uses described in Chapter 3.

Another purpose is to broaden our understanding of the science of chemistry. Chemistry is concerned with understanding the properties of substances, and their transformations. We are interested in all the chemistry that is possible, governed by natural laws, not just the chemistry that happens to exist in the world. We also want to put natural chemistry into context.

For example, natural DNA has a particular chemical structure, and it is not obvious that only that structure could have served its functions. Chemists have made analogs of DNA using sugars other than the one found in DNA to better understand what is special about the natur-

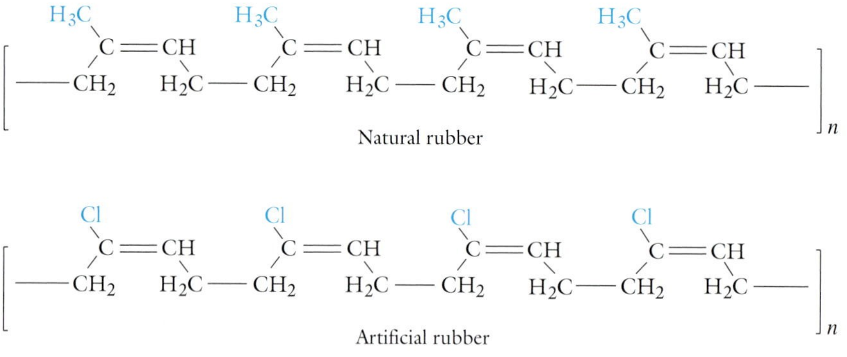

FIGURE 7.1

Top: a section of natural rubber, a polymer made up of linked five-carbon units. Four units are shown, but hundreds are actually present. *Bottom:* a section of artificial rubber, a synthetic polymer, in which the H₃C group of natural rubber is replaced by a chlorine atom.

al structure. They discovered that the properties of these modified DNAs show that they would not be as suitable for biological function as is the natural DNA.

Chemists frequently make analogs of known compounds. As another example, one kind of artificial rubber is an analog of natural rubber, in which the methyl group (H_3C) of natural rubber is replaced by a chlorine atom (Figure 7.1). The properties of the artificial compound are better for some uses.

There is a particular class of molecules that occur naturally but that can be studied only if we make them: the molecules that occur in space, but not on earth. There are giant clouds of unusual molecules whose collapse gives birth to the stars, and which are detected from the light they give off. From the nature of the light, chemists can speculate about the likely chemical structures in space. To confirm the speculation we make and study them in the laboratory.

Given these purposes, the chemist must still decide what to make. For potential medicines, analogy is a powerful principle, as discussed in Chapter 2. We make molecules that look like the substrates of enzymes or—even more effective—molecules that look like some of the unstable intermediates in an enzymatic reaction. We also examine the chemical structures of natural compounds with interesting biological effects to see how they might be improved.

Analogy has also driven the design of new useful materials, such as plastics and synthetic fibers. These are synthetic polymers. A polymer is a molecule made by linking many small molecules to form chains. Sometimes there are also links between the chains. Cotton is a polymer made up of thousands of linked sugar units. Proteins, including those in silk or in spider webs, are polymers made up of linked amino acid units. New polymers are made by chemists, by linking other small units together.

Wallace Carothers, at DuPont, got this field moving in the 1930s by creating nylon. He realized that silk, a protein, has what are called amide bonds in it, and he decided to make a polymer linked by amide bonds. However, he did not use natural amino acids, just simpler available chemicals. When he learned how to carry out the linking to make really large polymers, made up of thousands of units, he got nylon—a material that rivaled silk in many of its properties and was superior in some.

Sometimes new molecules are made to broaden or con-

95

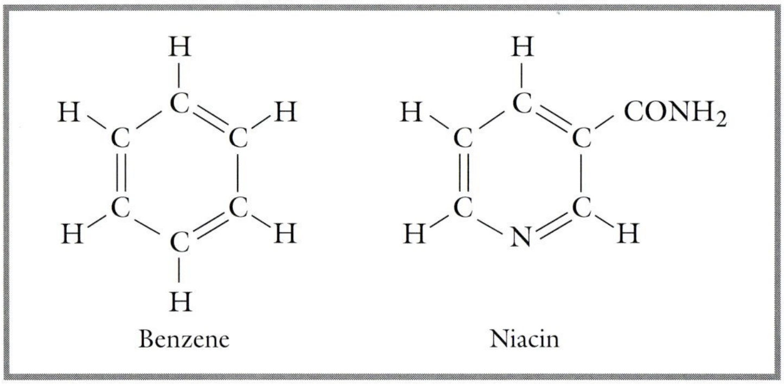

FIGURE 7.2

Chemical formulas of the molecule benzene (*left*) and the vitamin niacin (*right*). In both molecules the electrons in the ring can spread out, leading to unusual stability.

firm chemical theory. For example, many natural compounds contain what are called **benzene rings**, rings of six carbons with three double bonds and three single bonds connecting them (Figure 7.2). Benzene rings are part of the amino acids phenylalanine and tyrosine, for instance, which are present in almost every protein. Closely related rings are found in many vitamins, and in DNA and RNA. These rings are unusually stable, and chemical theory using quantum mechanics explained why they are so stable. Fundamentally, the reason is that some of the electrons in a benzene ring can spread out over the ring so as to diminish the repulsion of each electron by all the others.

However, the theory predicted that there would be many other rings that should also be especially stable, but they were not found in nature. The desire for molecules to test the theoretical predictions inspired chemists to synthesize new ring systems containing three carbons, four carbons, five carbons, seven carbons, and so on. The experimental result was that some of the new compounds were unusually stable, but not

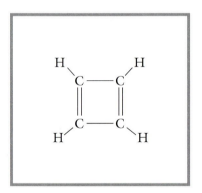

FIGURE 7.3

A molecule called cyclobutadiene. Although the electrons in the ring can also spread out as they do in the benzene ring, this molecule is not unusually stable but is remarkably unstable. At first this finding was a great surprise to chemists, but modern theory makes it clear why the six-carbon ring is stable while the four-carbon ring is not.

all of them (Figure 7.3). This finding was unexpected from the simplest theory about benzene, but was explained by more complex theories.

This synthetic program confirmed some aspects of chemical theory and caused a revision of some other aspects of the theory. It also made new types of rings available that had not existed before and that were incorporated into some useful compounds.

There are many examples of such programs in which new types of chemical structures were dreamed up and made by chemists. For example, compounds have been created in which a metal atom is held between two five-carbon rings like the meat in a sandwich. Such compounds—and in particular the open-faced sandwich version with a metal sitting on top of one five-carbon ring—have proven to be useful catalysts for synthesizing polymers. They have also extended our ideas of what is possible in the chemical world, since no such structures exist in nature.

How do chemists plan a successful synthesis of an unknown molecule ?

Analogy and imagination are the tools. Chemistry has an incredible literature describing past experiments, and all chemists expect to have to examine it to see what has already been done.

A good chemical library will contain journals—such as the *Journal of the American Chemical Society* or the journals of the British and the German chemical societies, for example—going back 100 years and more. A special publication from the American Chemical Society called *Chemical Abstracts* indexes and condenses what is in all these journals, so modern scientists can know where to look for publications related to their work. There are over 100 thick volumes of *Chemical Abstracts* just indexing chemical work of the 5-year period 1987–1991, taking up 26 feet of library shelf space. These are just the indexes; the condensed accounts of each chemical article are far more voluminous. Luckily, modern computers have arrived just in time to let all this information become available electronically.

Chemists wanting to make an unknown molecule will first see how related compounds were made. Then they will

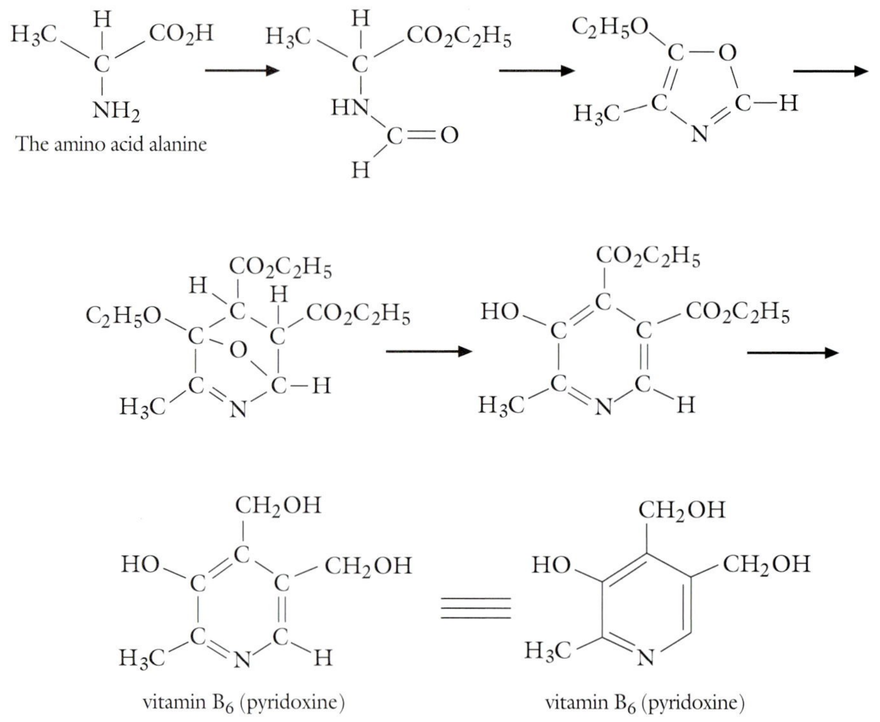

FIGURE 7.4

A synthesis of vitamin B_6. The details of the chemical reactions involved in each step are not shown, since they require a level of chemical understanding above that assumed for this book. However, the reader will see that the final product does not look much like the starting material, an available amino acid. The creative insight was that the six-atom ring in the final target could be made from the five-atom ring in the third structure. The final vitamin is shown in two forms. In the one on the left we draw in all the atoms, while in the structure written on the right we use a common chemical simplification in which carbon atoms are assumed but not shown at the angles of the ring, and any attached hydrogen atom to such an invisible carbon is also not shown.

consider whether they can use a similar route, guided by chemical theory that tells them how similar the new chemistry is expected to be. This strategy is planning by analogy.

Sometimes there are no good analogies, especially if completely new types of molecules are the target. Then chemists use their imaginations, guided again by chemical theory. What is done is a little like what is done by good chess players.

First chemists consider the "end game". What final

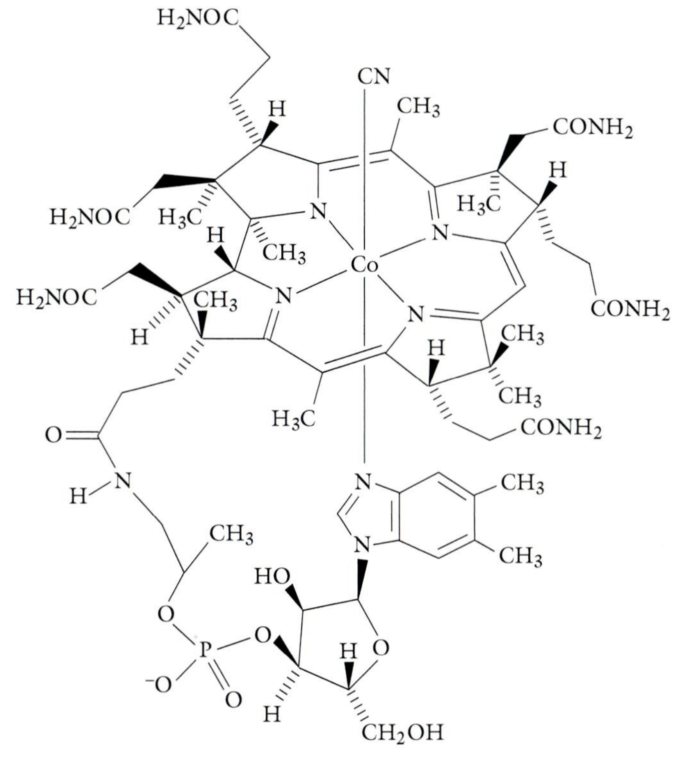

Vitamin B$_{12}$

FIGURE 7.5

In a monumental feat, vitamin B$_{12}$ was synthesized by a research group under the direction of Robert Woodward in collaboration with Albert Eschenmoser. Woodward received the Nobel Prize in Chemistry in 1965 for such achievements. The starting materials were simple available chemicals, as in the vitamin B$_6$ synthesis, but many more steps were required and new chemical ideas were needed. The structure is shown with the simplification common for such large molecules—every angle that has no atom shown has a carbon atom, and each carbon atom carries as many unshown hydrogens as are needed to bring the total number of bonds at carbon to four. In this synthesis the geometry had to be controlled so that the correct bonds were either up or down. There is no such requirement in vitamin B$_6$.

steps would lead to the target from simpler more available compounds? Does theory suggest that those steps would work in this particular case? Can one invent a new chemical reaction that should work, again guided by theory? Then they consider the "middle game". How can one make the precursors from even simpler compounds? Finally, the "opening". What initial chemical reactions could be applied to commercially available chemicals to get started along the correct path? Of course the experiments are actually done in the reverse order.

A sequence of as many as 30 steps may go into such a plan for a complicated target. The difference between this plan and a 30-step chess plan is that we don't have a thinking opponent, whose responses are not predictable. However, nature may behave unpredictably if our theory is not solid enough, so the plan may not work. If it does not, a new plan is devised to take account of what has been learned. As in chess, we may have alternative versions of what we will do at step 15, depending on what happens at step 14 where a risky reaction is being tried. Figure 7.4 shows a commercial synthesis of a vitamin, and Figure 7.5 shows one of the most complex molecules that has yet been synthesized from simple starting materials by chemists.

Where is this work done?

Much of the work is done in university laboratories by the chemistry faculty assisted by graduate students (who are working toward master's or doctoral degrees) and postdoctoral students who already have Ph.D. degrees. Some of the work is done in government laboratories, such as those at the National Institutes of Health. In addition, the synthesis of new chemicals is done in the large number of research laboratories throughout industry. Medicinal chemists in the pharmaceutical industry are particularly active in synthesis. As many as 10,000 new chemicals must be created, and tested, before one chemical passes all the tests and becomes a commercially available medicine.

Medicinal chemists use the most modern advances in chemical basic knowledge to make new compounds for biological testing, or to devise highly efficient ways to manufacture them once they have been found to have interesting biological activity (Figure 7.6). Researchers in the chemical industry make products such as new plastics, new dyes, new paints, and new insecticides, for example, with the same combination of analogy and imagination. The high level of training that the Ph.D. degree represents is valuable even for the two-thirds of Ph.D. chemists who go into industri-

FIGURE 7.6

The synthesis of new chemicals involves a combination of experimental work and computer design

al research. Their advanced training lets them excel in a highly competitive world.

What is done with a new chemical once it has been made ?

For a chemical intended for practical application there are always two questions that must be answered: Does it have the desired useful properties? Does it have harmful side effects?

If tests indicate that the answers are as hoped for, the new compound will probably be patented. A patent gives the inventor a few years in which to develop the practical applications, and sell the resulting products, without competition from others who have not done the hard work of research that led to the discovery.

Some countries experimented with patent laws that did not permit the patenting of new medicinal compounds. The theory was that sick people should benefit from a new medicine at the cheapest possible price, which competition would ensure. Unfortunately, the result was that no one was willing to do medicinal chemistry research in those countries, since the discoverer would not even recover the cost of research if there were no patent protection. In most countries, the exemption of medicines from patent protection has now been dropped.

Research results that have been protected by patents, or that are not patentable, are published in the chemical journals. Chemical journals also publish research aimed at purely theoretical questions, such as much academic research. The synthesis itself might be sufficiently interesting to be published, or it might be accompanied by accounts of the properties of the newly created chemicals.

The useful properties of new molecules are not always obvious to the original creator of the compounds. For instance, the metal "sandwich" compounds described earlier were made purely out of scientific curiosity about a completely new type of structure. This curiosity led to the Nobel Prize in Chemistry for Geoffrey Wilkinson and Ernst Fischer in 1973. Only later did others realize and demonstrate that some of the compounds could be useful catalysts for making polymers.

T H E   F U T U R E

The targets of synthesis are limited only by chemists' imaginations. However, it is possible to describe some areas of current and future interest.

For one, synthetic chemists are trying to devise practical ways to make some important chemicals that are rare. A good example is a medicinal chemical isolated from yew trees. This substance has important anticancer properties, but it is scarce enough that chemists want to make it by synthesis. However, its chemical structure is a challenge for our current synthetic methods. It has been made in a laboratory, but not yet with a procedure that seems practical for manufacture from very simple available chemicals.

Complex carbohydrates are also targets. They play an important biological role as the labels on the surfaces of cells. For example, they are the labels that distinguish type A blood from type B. They distinguish normal human cells from those of invading bacteria. There is also evidence that bacteria bind to carbohydrates on the surfaces of human cells in order to infect them. Thus such carbohydrates, as free molecules, should be able to bind to the bacteria and prevent them from attacking cells. Complex carbohydrates could play useful roles in medicinal chemistry, but good methods are still needed to prepare them easily.

Synthesis will also be used to pursue some of the goals described in earlier chapters. New chemicals are needed that will let us dispose of radioactive waste from nuclear power plants, and new catalysts are needed to let us perform chemical reactions with the achievement of selectivity like that common in biochemistry. Here nature does not furnish the target, as with the cell-surface carbohydrates. Chemists must design and build new compounds that they believe will have useful or interesting properties. The challenge is picking the right targets, as well as figuring out how to make them.

In the more basic area, there is a continuing effort to make new chemicals with structures very different from those in the natural world, to see what unusual properties the new structures might have.

There is a lot of interest in developing new general synthetic methods that may be widely used in making various target molecules. Particular efforts involve devising chemical reactions that produce only the desired product, with no by-products requiring disposal. It is especially important to devise good methods to make the desired product with the correct handedness, especially for new medicinal compounds. It is often found that the left-handed isomer, for instance, is a useful drug while the right-handed isomer has undesirable properties. Most chemical reactions produce an equal mixture of left-handed and right-handed molecules, which must be separated. New synthetic methods that imitate nature's ability to produce only one isomer are still being developed.

There is a concern to perform chemical reactions in water if possible, in order to avoid other solvents that may be environmentally harmful. There is a desire to have highly efficient syntheses. They should have as few steps as possible, and each step should proceed in very high chemical yield, with few side products. They should also be efficient with respect to energy and materials.

103

A particularly interesting development, with more to come in the future, is the use of computers to assist in synthesis planning. In some approaches the computer is taught to think the way a good synthetic chemist would think; in others the computer simply evaluates huge numbers of known reactions to see which ones might best be put together to get to the target. Related to this computer planning is the desire to develop computer-controlled robotic systems to carry out synthetic reactions in the laboratory.

Finally, there is much interest in what are called biomimetic methods. Nature makes a number of complex molecules efficiently. Chemists are trying to imitate the pathways used and are also trying to imitate the enzymes that catalyze biochemical reactions. It seems likely that selective catalysts that imitate enzymes will play an increasing role in chemical synthesis.

Understanding Molecular Structures and Chemical Change

> All that science can achieve is a perfect knowledge and a perfect understanding....
>
> —H. Helmholtz, *Academic Discourse*

When a new chemical is isolated from nature, or created in the laboratory, its detailed molecular structure has to be determined. When a chemical reaction occurs, in the laboratory or in a living cell, we want to understand what has happened and how it took place. In this chapter we will describe the questions asked in these areas of chemistry, and the methods currently used to answer them.

How do chemists determine the structures of new molecules ?

The first job is to show that the new substance is a pure chemical. The usual test is to show that it cannot be separated into other chemicals. Modern separation methods are so powerful that they separate virtually any mixture. Then the chemical composition is determined—that is, how many atoms of carbon, hydrogen, oxygen, or iron, for example, does it contain.

If the molecule is not enormous, its composition is normally determined now using a mass spectrometer, which determines the weight of a molecule of the new substance (Figure 8.1). From this weight, the chemical composition can be determined. An oxygen atom (atomic weight = 16 units) can even be distinguished from the nominally same weight of a carbon atom (atomic weight = 12 units) plus four hydrogen atoms (atomic weight of each = 1 unit) since the precise masses of these atoms are slightly different from the numbers given. For example, by convention the atomic weight of carbon is assigned the value 12.000000, but based on that assignment the true atomic weight of oxygen-16 is 15.995915 and that of hydrogen is 1.007825. Thus a CH_4 unit has a weight of 16.0313, different from the 15.995915 of oxygen. With modern instruments the tiny differences can be detected, so the true molecular formula can be determined.

It is one thing to know which atoms are in a molecule, but another thing to know how the atoms are connected.

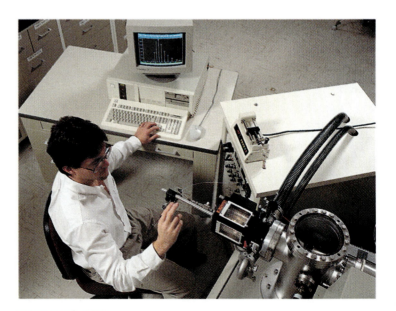

FIGURE 8.1

A modern mass spectrometer, used to determine the composition of new molecules.

In a mass spectrometer the molecules can break up into pieces. The composition of the pieces, determined from their exact weight, tells a lot about what the connections are. Other methods to determine the atomic connections are also very powerful.

Consider again the two isomers with formula C_2H_6O that were described in Chapter 1. They can be distinguished using nuclear magnetic resonance (NMR).

Ethyl alcohol   Dimethyl ether

An NMR spectrum reveals how many different types of hydrogens a molecule contains. The six hydrogens in dimethyl ether are all equivalent, all attached to the same type of carbon atom. Thus dimethyl ether shows only a single signal in its hydrogen NMR spectrum. However, in ethyl alcohol there are three hydrogens on the end carbon, two hydrogens on a different type of carbon (that carries an oxygen atom), and one hydrogen on the oxygen atom. Its hydrogen NMR spectrum shows one peak for the OH hydrogen, another of twice the size for the CH_2 hydrogens, and another of three times the size for the CH_3 hydrogens.

The further details of this spectrum give information about the chemical environment of these hydrogens. Where the various signals appear in the spectrum differs because of the electron-attracting properties of the oxygen atom, while the detailed structure of each peak tells about the number of hydrogens on *neighboring atoms*. For instance, the peak for the CH_3 group has a shape indicating that there are two hydrogens on a neighboring carbon atom.

NMR can also be used to determine how many different types of carbons a molecule contains. Dimethyl ether has one type of carbon, so only one signal appears in the carbon NMR spectrum. Ethyl alcohol has two different signals in the carbon NMR spectrum. In addition, the details of a carbon NMR spectrum reveal how many hydrogens are on each type of carbon atom, and even how many hydrogens are nearby. Using all the power of NMR spectroscopy, it is normally possible to deduce the detailed atomic arrangements in simple chemical compounds, and even the three-dimensional structure of small proteins that contain hundreds of atoms.

107

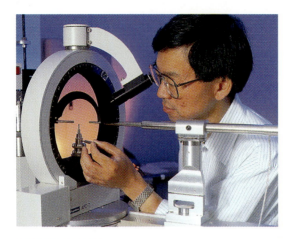

FIGURE 8.2

In the X-ray determination of a molecular structure, a tiny crystal is mounted in the X-ray beam.

The other very powerful technique for structure determination is X-ray crystallography (Figure 8.2). As the name implies, X-rays are beamed onto solid crystals of an unknown compound. The X-rays spread out when they pass through the crystal, like water passing through a spray nozzle. Computer methods can translate the pattern of such a spray into the chemical structure whose atoms caused the spray as the X-rays bounced off them. For example, the structure of vitamin B_{12} shown in Chapter 7 was determined by Dorothy Hodgkin using X-ray crystallography. She received the Nobel Prize in Chemistry in 1964 for this structure determination and related work.

Many other instruments also contribute. An infrared spectrometer and an ultraviolet or visible light spectrometer can detect the presence of some particular chemical arrangements in a new molecule. Other instruments can determine the "handedness" of molecules. Modern instruments have greatly simplified the problem of determining structures.

What is the relationship between the detailed structure of a molecule and its properties?

This is one of the central questions of chemistry, and it is not always easy to answer. For example, the project to sequence human genes (described in Chapter 2) will lead to information about the sequences of amino acids in the proteins coded for by those genes. Then computer methods will eventually be able to predict the likely three-dimensional structure of one of the proteins, the way in which it will fold (discussed in Chapter 5). But will we be able to look at that structure and say what the biological role of the protein is, based on its chemical properties? Perhaps, but only after we learn some more basic chemistry.

Luckily, there are families of chemicals that have similar properties. For example, Chapter 1 showed the structure of dimethyl ether, and Chapter 6 described diethyl ether. In dimethyl ether an oxygen atom joined two one-carbon groups, and in diethyl ether an oxygen joined two two-carbon groups. To a first approximation the chemical properties of the two ethers are similar, as are those of other ethers, compounds in which an oxygen atom links two carbon groups.

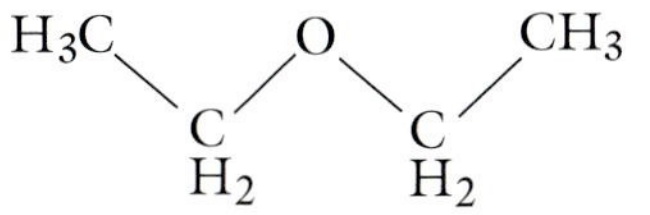

Diethyl ether 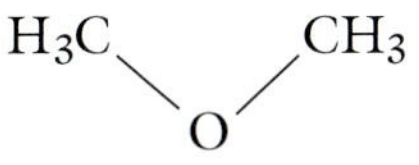   Dimethyl ether

The chemical formulas for ethyl alcohol and for methyl alcohol are shown below. They have similar properties, which are different from the properties of the ethers. In the class of alcohols, an oxygen atom is linked to one hydrogen and one carbon group.

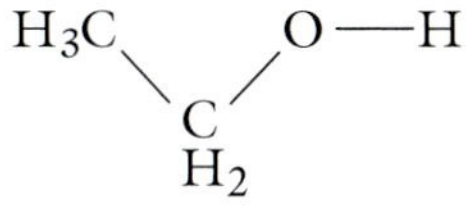

Ethyl alcohol 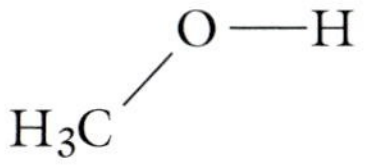   Methyl alcohol

Within a chemical family, there are differences as well as similarities. With experience, we have learned how a methyl group differs from an ethyl group in its chemistry. With modern computer simulations of reaction pathways, it is sometimes possible to account for differences of this type in terms of the detailed geometries of the molecules.

Another factor that causes differences within a family of related compounds is electronic. For example, in the conversion of ethyl alcohol to diethyl ether described in Chapter 6, a step was shown in which a proton added to the oxygen atom to put a positive charge on the ethyl alcohol.

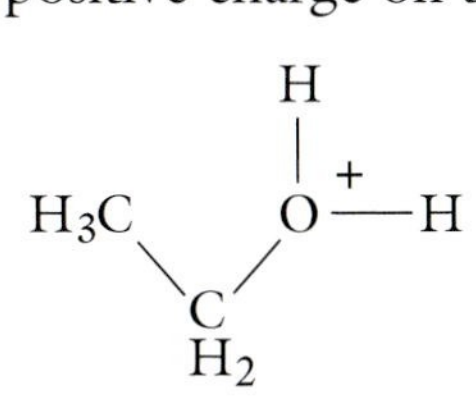

Ethyl alcohol with a proton
added to the oxygen atom.

If the carbon groups carry not hydrogen atoms but other atoms (such as fluorine) that make nearby positive charges unstable, it is harder to put a proton on the OH group. Those alcohols with attached fluorine atoms do not react as rapidly to form ethers.

To make sense of such effects, it is critical to know the pathways involved in chemical reactions. For instance, we needed to know that a positive charge was put on the oxygen atom of ethyl alcohol to understand why fluorine atoms make the reaction more difficult (slower). This leads to the next question.

How do chemists determine the detailed pathways by which chemical reactions occur

The methods of study depend on how large and complicated the reacting species are. At the simplest end of the scale, chemists study what happens when a hydrogen atom (H) flies through empty space and collides with a chlorine molecule (Cl_2), in which two chlorine atoms are linked. The result is the formation of a molecule of hydrogen chloride (H—Cl), with a bond between the hydrogen and one chlorine atom, while the other chlorine atom (Cl) flies off (Figure 8.3).

In such a reaction, the questions raised include the following: (1) Does the hydrogen atom have to collide directly with the chlorine molecule for a reaction to occur? (2) If so, what is the best angle of collision? Does the hydrogen atom collide with only one of the chlorine atoms, perhaps hitting the Cl_2 molecule end on, or does it prefer to hit the

$$H \; + \; Cl — Cl \; \longrightarrow \; H — Cl \; + \; Cl$$

FIGURE 8.3

Collision of a hydrogen atom with a chlorine molecule produces a molecule of H—Cl and a chlorine atom. What is the best direction of approach of such a hydrogen atom, and where does the energy end up after the reaction?

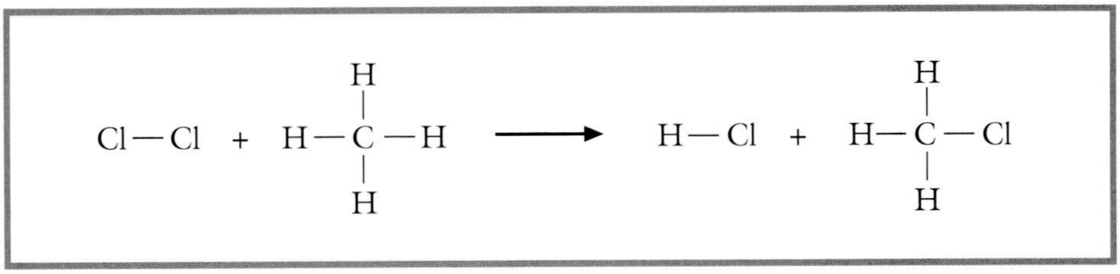

FIGURE 8.4

The reaction of a chlorine molecule with a methane molecule to form hydrogen chloride and methyl chloride. The reaction does not occur unless light is shined on the mixture. This is a clue that indicates how the reaction occurs.

middle of the Cl_2 molecule in order to split the Cl—Cl bond? (3) How hard does the collision have to be (how energetic) for the reaction to occur? (4) The products of the reaction have some extra energy. Where is that energy found? Is it found in the speed by which the chlorine atom flies off? Is it in vibrations of the new H—Cl molecule?

The methods used to study such questions are very close to physics. Physical chemists use beams of atoms or molecules aimed at each other in large evacuated chambers, perhaps using light from a laser as well. The pioneering work establishing these methods was recognized with Nobel Prizes to chemists Dudley Herschbach, Yuan Lee, and John Polanyi in 1986. In these studies the angles at which atoms and molecules collide can be controlled, and the angles at which the products fly off, and their energies, can be measured. Thus direct evidence is available to answer the questions asked.

On a larger scale, chemists have investigated the reaction of Cl_2 with methane (CH_4). A molecule of H—Cl is formed, and the other product is methyl chloride, in which there is a new carbon–chlorine bond (Figure 8.4). Critical in such studies is the use of what is called a controlled experiment. That is, the same experiment is repeated while the chemist changes one factor at a time, to see whether it is important. For example, even though the reaction in Figure 8.4 is favorable—the products are more stable than the starting materials—nothing happens when Cl_2 gas and methane gas are simply mixed in the dark. However, when visible light is shined on the mixture a rapid reaction takes place to form H—Cl and methyl chloride. What is happening?

111

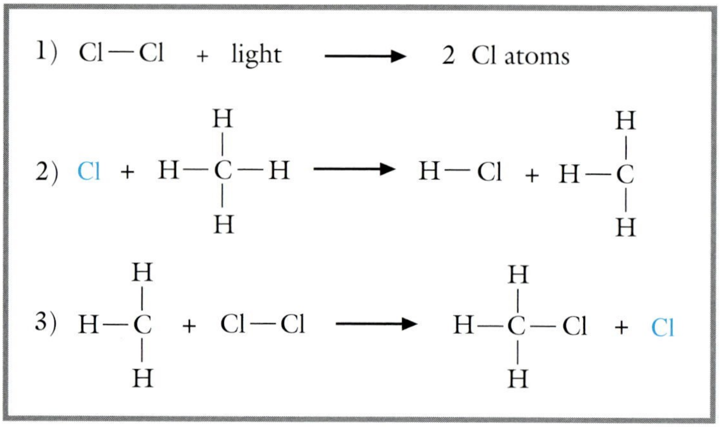

FIGURE 8.5

The pathway by which Cl_2 and methane react to form H—Cl and methyl chloride. In step 1 the light is absorbed by Cl_2 and breaks it into chlorine atoms. In step 2 a chlorine atom removes a hydrogen atom from methane, forming H—Cl and CH_3. In step 3 the CH_3 reacts with Cl_2 to form methyl chloride and a new chlorine atom. This chlorine atom can then perform step 2 again. Thus step 2 and step 3 happen over and over in a chain reaction. Only a little light is needed to start the chain, in step 1.

112

Cl_2 is a green gas, but methane is colorless. Thus the light can only be absorbed by Cl_2, since visible light passes right through colorless materials and will have no effect unless it is absorbed. In other experiments it is found that the most effective color of light is exactly that which is best absorbed by Cl_2. When the light energy is taken up, the Cl_2 molecule breaks into two Cl atoms. A chlorine atom then takes a hydrogen from methane, making H—Cl. After the hydrogen atom is lost, methane becomes CH_3, and this species then takes a chlorine atom from Cl_2 to make the methyl chloride product. The remaining Cl then attacks methane again, and the process repeats itself.

This mechanism (Figure 8.5) is what is called a chain reaction, in which the light just starts it, but then each chemical step generates a reactive species—a chlorine atom (Cl) or a methyl group (CH_3)—that can rapidly react with a methane or chlorine molecule. The evidence for this pathway is that no reaction occurs in the absence of light, and that a single unit of light called a photon can start a chain with as many as 1000 following reactions before a side reaction stops the chain. That is, the light energy is not used for every step; it is just needed to get the whole chain started.

In this kind of study the pathway can be determined with high probability, but not with the detail seen in the simple reaction we first described. With the reactions of still larger molecules, the evidence becomes even more circumstantial and indirect. The study of how the rate of a reaction depends on concentrations and on reaction conditions is a very common tool in the study of reaction pathways. However, it is an indirect tool, and sometimes leads to ambiguous pictures.

For example, the enzyme ribonuclease A consists of a coiled-up chain of 124 linked amino acids. It binds to RNA and catalyzes the cleavage of the RNA chain. In spite of much work, the pathway by which this reaction occurs is still in dispute.

One important piece of evidence comes from studying how fast the cleavage reaction goes as one increases the concentration of the RNA. With simple reactions that don't involve enzymes, the speed of the reaction goes up as the concentration of the reacting molecules goes up; there are more collisions that can lead to reaction when more molecules are bouncing around in a given volume. However, with the ribonuclease–RNA reaction, and indeed with essentially all reactions in which enzymes are catalysts, the speed of the reaction increases up to a certain point as the concentration of the reacting molecule increases, but then it levels off.

The leveling of the speed occurs because the RNA binds to the enzyme, and the reaction occurs within the resulting complex of enzyme with RNA. When the concentration of RNA is sufficient to bind to all the enzyme, adding more RNA will not increase the rate. We say that the enzyme is saturated with RNA. This was the earliest evidence that enzymes bind to the molecules whose reactions they catalyze. It has recently been confirmed by isolating the complexes and determining their structures by X-ray methods.

A second piece of evidence is that the rate of the RNA reaction is greatest when the solution is not acidic or basic, but is neutral. From the dependence of the rate on the acidity of the solution, it is clear that the enzyme uses two of its amino acids in a special way. The amino acid histidine can be changed from its basic form to an acidic form when it is in solutions containing acid, and the detailed effect of solu-

113

tion acidity on rate indicates that two histidines play an important role in the catalysis. The enzyme consists of 124 amino acids linked in a coiled chain; the histidines at positions 12 and 119 (counting from one end of the chain, so they are at almost the opposite ends of the 124-unit chain) are critical and the coiling of the chain brings them close together.

The enzyme is active when one of the histidines exists in its basic form while the other one is in its acidic form. For the cleavage of RNA, other chemical studies show that a catalyst will work best if it has both an acid and a base group in it. Furthermore, the chemical structure of the complex between RNA and the enzyme, determined by X-ray, shows that the two histidines are right next to the spot where RNA cleaves. Thus the general outline of the mechanism is clear: RNA binds to the enzyme, and then the two histidines of the enzyme cooperate in cleaving the RNA, one in its basic form and the other in its acidic form.

There is still disagreement about the precise way in which these two histidines cooperate to cleave the RNA. At least two different paths are possible, consistent with other chemical experience and with the evidence about the enzyme. The problem remains the subject of active research, and it is important. We need to know how enzymes catalyze their reactions to understand why the chemistry of life is so effective. Also, when we understand the way they perform such effective catalysis, we can design other catalysts to imitate them, and medicines to interact with them.

As we will discuss, the future of elucidating reaction pathways will surely require different tools, currently being developed.

Where is the work on molecular structures and properties, and on reaction pathways, done ?

This work is largely done in universities, by research teams consisting of faculty and students. Commonly a chemistry professor directs the research of a group of students—perhaps a few undergraduate students, some graduate students working for Master's or Ph.D. degrees, and perhaps a few postdoctoral students who already have Ph.D. degrees but are getting additional training.

The professor will normally think up the design of the experiments and suggest how they should be done, but the actual experiments are done by the students. By analogy, the professor is the architect and contractor, while the students actually build the house. In the course of their work they are learning how to perform frontline research. Afterward they will make use of this training if they themselves become professors somewhere, or more likely when they take up careers in the chemical or pharmaceutical industries.

Work on the pathways used by enzyme catalysts was traditionally done in biochemistry departments, but it is increasingly part of the research programs of university chemistry departments. Medicinal chemists make great use of the information about how enzymes work when they design new drugs. However, they normally depend on the results from university laboratories rather than doing pathway studies themselves.

 T H E F U T U R E

In these areas of chemistry, progress will require the creation of better tools and better theories.

Structure Determination

X-ray methods for structure determination are now so powerful that the chief limitation is technical—the methods can be applied only to crystals, and not everything forms nice crystals as salt and sugar do, not even every solid. Also, the detailed chemical structure that X-ray methods determine is that of the molecules in their crystalline forms, and they may have different shapes when they are dissolved.

As a simple example, the sodium atoms and the chlorine atoms of sodium chloride (common salt) are next to each other in the crystals, but in water solution they float apart. As a more common worry, proteins are often flexible, and the shapes they take up to pack into a crystal can be different from the shapes they relax into in water solution. For these reasons, among others, there is great interest in using nuclear magnetic resonance (NMR) methods to study the solution structure of proteins, and to relate this structure to the one that X-ray methods detect in the crystal. Also, NMR can be applied to molecules that don't form crystals.

Another challenge is to determine the structure of natural materials that are available only in very tiny amounts, too little to form crystals that can be examined by X-ray methods. Some natural compounds with important biological activity are in this category, and determining their structures represents a challenge for the future. However, in general the current methods are suitable for most structure problems. Furthermore, some new types of microscopy are being developed that hold the promise of permitting a direct look at individual molecules and their structures. Thus the future of this field will consist of a series of discoveries of biologically important new substances, and the determination of their structures by modern methods.

Reaction Paths

The tools are more of a problem in this field. Most chemists working to discover reaction paths agree that an important eventual tool will be the computer. That is, in the future we will have theories and computer methods with the ability to describe exactly how any chemical reaction occurs. They will be able to generate a movie showing atoms moving in the course of a reaction, indicating the speed of the various steps and the energy changes associated with them.

It is possible to do this now, but not with the needed reliability. The theories have to be simplified to allow even very fast computers to handle them in a feasible time, and simplifications produce errors.

Even in the reactions of small molecules, experiments will be needed for quite a while to establish the key points of a reaction path. For example, does the oxygen atom in a product come from the oxygen atom of water (H_2O) or an oxygen atom from the air (O_2)? Oxygen isotopes can be used to answer this question, using special O_2 containing oxygen-18 instead of the oxygen-16 that is in normal O_2 or water, and seeing what kind of O ends up in the product. The oxygen isotopes oxygen-16 and oxygen-18 contain the same number of protons in the nucleus, eight, but oxygen-18 has ten neutrons while oxygen-16 has only eight. They can be distinguished by the use of a mass spectrometer. However, none of the current experimental methods permit the determination of a reaction path in full detail. The details are still added to the key points with a combination of imagination and insight.

When more atoms are involved, obtaining a reliable computer calculation becomes more difficult. Thus it will be a while before we can trust theory to predict the paths of enzyme reactions reliably. Since the reactions occur in water or in biological membranes, the computer calculation must also include the surrounding atoms in addition to those of the enzyme and the reacting molecules. For this reason there is a lot of excitement about the possibility that X-ray methods might be able to contribute here.

Since X-rays take a "picture" of the molecule, why not just use X-ray movies? The methods are currently not fast enough—it can take hours to get the data for an X-ray structure determination, while chemical reactions go on in tiny fractions of a second. Also, the X-ray methods work only with crystals.

Do chemical reactions occur in crystals? Many reactions do, and in particular many enzymes are still active catalysts even when they are present as crystalline solids, not in solution. Currently some enzyme crystals have been cooled to low temperatures, to slow down the reactions, and X-ray methods have determined the structures of a few intermediate states along the reaction path. It's not the same as a detailed movie, but it is a promising indication of future progress.

Some very fast physical methods are now able to operate on the extremely short time scale of a chemical reaction. Progress is being made on getting X-ray methods fast enough to make the idea of X-ray movies seem less a dream, although still not a reality. In addition, modern laser techniques allow chemists to use a laser pulse to start a chemical reaction, then in a femtosecond or so (a femtosecond is 0.000000000000001 seconds) to fire in another laser pulse to get the spectrum of a reaction intermediate (Figure 8.6). With such techniques it is possible to "see" a simple chemical reaction. In the future we may be able to apply these techniques to more complex reactions, including enzyme catalysis.

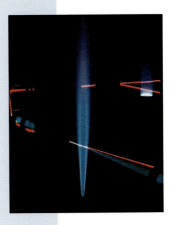

FIGURE 8.6

A flame is the result of chemical reactions. Here two laser beams are being used to determine what molecular species are present in the flame, in order to understand the chemistry in detail. Such studies have led to improved diesel engines, which produce less soot.

117

POSTSCRIPT

119

In this book I have tried to give the reader some idea of the great range of chemistry and its importance. In common with other scientists, chemists explore the natural world and try to understand it. However, chemists also expand the world, creating new chemicals and inventing new chemical reactions.

There are many important questions waiting to be answered, and many practical goals that will permit chemistry to contribute even further to human welfare. Students considering chemistry as a career have a choice of creating new medicines or making other contributions to human health, devising new materials with important applications, helping to improve the environment, helping to advance the computer revolution, helping to elucidate the fundamental chemistry of life, or helping to understand the fundamentals of chemistry itself.

It is an important challenge. I hope that some readers will be inspired to take up the challenge, and make their own contributions to the science that is uniquely central, useful, and creative.

FURTHER READINGS

Several books will lead the reader further into the areas I have discussed.

1. Ball, Philip. *Designing the Molecular World: Chemistry at the Frontier,* Princeton University Press: Princeton, NJ, 1994; 376 pp. This book addresses many of the questions raised in my book, but at a level requiring more scientific sophistication. It includes a very extensive bibliography that can guide readers further into the details of the topics considered. A weakness is a tendency to devote considerable attention to some fashionable current work that may not stand the test of time.

2. Gray, Harry B.; Simon, John D.; Trogler, William C. *Braving the Elements;* University Science Books: Sausalito, CA, 1995; 418 pp. This excellent book is aimed at an audience with little or no past education in chemistry, and explains the amount of chemistry needed to understand a number of ways in which chemistry affects our lives. The writing style is exceptionally user friendly.

3. Hoffmann, Roald. *The Same and Not the Same;* Columbia University Press: New York, 1995; 294 pp. A very readable account of some topics in chemistry that relates them to the artistic world. The intellectual beauty of the field is particularly well conveyed.

4. James, Laylin K., Ed. *Nobel Laureates in Chemistry, 1901–1992;* American Chemical Society: Washington, DC, 1993; 798 pp. A description of the lives of all the winners, and of the work that led to their Nobel Prizes. It is not very technical, and easily read by those with little background in chemistry.

5. Free, Helen. *Your Chemical World;* American Chemical Society: Washington, DC, 1994; 84 pp. A highly readable account of some of the ways in which the world is made better by the products of chemistry.

6. Burger, Alfred. *Understanding Medications;* American Chemical Society: Washington, DC, 1995; 206 pp. A larger discussion of the topics of my Chapter 2.

7. Good, Mary L., Ed. *Biotechnology and Materials Science: Chemistry for the Future;* American Chemical Society: Washington, DC, 1988, 135 pp. Predictions of the future developments in two important areas in which chemistry contributes, in chapters written by experts in the fields.

8. Alper, Joseph; Nelson, Gordon L. *Polymeric Materials: Chemistry for the Future;* American Chemical Society: Washington, DC, 1989; 110 pp. More information about the new materials that chemistry has created and that are discussed in my Chapter 3.

9. Pimentel, George C. *Opportunities in Chemistry;* National Academy Press: Washington, DC, 1985; 344 pp. A classic, if a little old now. For those interested in a career in chemistry, it points out many of the discoveries still to be made and what their impact will be. It will also be of interest to those who want to know what chemistry may contribute to all our futures.

10. Hegedus, L. Louis. *Critical Technologies: The Role of Chemistry and Chemical Engineering;* National Academy Press: Washington, DC, 1992; 70 pp. A concise summary of the importance of chemistry and chemical engineering to the economy, the security, and the well being of the United States. The book is particularly well-illustrated.

11. "What's Happening in Chemistry"; Department of Media Relations/Local Section Public Outreach, American Chemical Society: Washington, DC, annual. These compact reports describe advances of the preceding year in nontechnical language. They convey some of the excitement that accompanies front line research, and the benefits it creates.

122

INDEX

123

Catalysts
  basic definition, 10, 77
  challenges for future chemists, 89–91
  chemical description, 78–80f
  destruction of ozone by chlorine, 60–61f
  enzymatic, 79–85
  importance in chemical reactions, 9
  solid metal catalysis, 86–87
  See also Enzyme entries, Reaction
      pathways
Catalytic converters
  environmental purpose, 39, 56, 78
  role of rhodium in catalytic
      hydrogenation, 87
Catalytic hydrogenation, 86–87f
CDs, *See* Compact discs
Cells
  chemistry of living and artificial systems,
      32
  See also Fuel cells
Cellulose
  chemistry of wood, 2
  removal of lignin to make paper, 44–45
Ceramics
  chemistry of composites, 47
  as possible superconductors, 48–49
  use in home fixtures, 35
CFCs, *See* Chlorofluorocarbons
Chain reaction, description and example,
      110–113
Chaperone proteins, role in pathway of
      protein folding, 73
Charcoal, early chemistry, 2
Chemical Abstracts, description, 97
Chemical change, definition, 2
Chemical codes
  human genome project, 27–28
  insect mating, 59
Chemical collisions, 110–111
Chemical engineering, description, 11
Chemical industry
  healthy balance of trade, 14
  overview, 10–14t
  See also Chemical process industries
Chemical literature, 97
Chemical Manufacturers Association,
      program for Responsible Care, 53–54f
Chemical output, U.S. share of worldwide
      output, 13f
Chemical process industries
  description and economic impact, 12f–14
  importance in airplane and auto
      manufacturing, 39f–40

Chemical process industries—*Continued*
  role in home construction, 35
Chemical properties
  fundamentals, 7–9
  predictions with computational chemistry,
      76
  relationship to chemical structure,
      108–110
Chemical reactions
  details studied by chemists, 11
  example of photosynthesis, 25–27
  formation of ether from alcohol, 78–80f
  fundamental principles, 9
  prediction of mechanism with
      computational chemistry, 76
  role of catalysts, 9
  time scale, 117
  See also Binding studies, Enzyme entries,
      Explosives, Reaction pathways
Chemical sensors, description and use in
      electronics, 74–75
Chemical structure
  determination with computational
      chemistry, 75
  HIV enzyme, 22f
  protein ribbon diagram, 72f
  relationship to chemical properties,
      108–110
  See also Enzyme–substrate complex,
      Three-dimensional chemical
      structure, X-ray entries
Chemical warfare, description, 42
Chemistry
  central questions, 4
  earliest, 2–4, 31–32
  fundamental principles, 7–9
  general definition, 1
  in everyday life, 33–45
  role of alchemy, 6f
  See also Biochemistry, Chemical process
      industries, Medicinal chemistry,
      Synthetic chemistry
Chemists
  collaborative work, 67–70, 73–75
  education, 100–101, 114–115
  primary activities, 2
  See also Future challenges for chemists
Chirality
  description, 8–9
  See also Handedness
Chlorine atoms
  in artificial rubber, 94f
  role in ozone destruction, 60–61f

Drugs—*Continued*
   sulfa drugs, 18–20
   See also AIDS, antibiotics, Medicinal
      chemicals, Pharmaceutical industry
Dyes
   color photography, 45, 48
   in everyday life, 34
   inks in books, 44
   natural sources, 2
   sulfonamides, 18–19

E

Eastman Chemical Company, landmark
      environmental features, 54
Educational requirements in chemistry
   chemists, 100–101, 114–115
   military school graduates, 42
   physicians, 28
Einstein, Albert, cautionary quotation, 93
Electric cars
   prototype, 56*f*
   research interest of auto industry, 39–40
   search for effective battery design, 57–58
   See also Automobiles
Electrical engineers, collaborative work with
      chemists, 67–70, 73–75
Electricity
   future hydrogen economy, 58
   future role of superconductors, 48–49
   role of chemical products, 36
Electronic noses and tastebuds, future
      chemical advances, 75
Electronics
   contributions of chemistry, 67–70
   potential future advances, 73–75
Energy states
   allowable protein geometries, 71–73
   relation to likelihood of reaction, 9
   See also Catalysts, Enzyme entries
Entropy
   chemical definition, 9
   relation to likelihood of chemical reaction,
      9
Environmental chemistry
   concept of limited stability, 64–65
   future challenges, 64–65
   general description, 10, 53
   industrial chemicals that mimic female
      hormones, 63
   positive example of Eastman Chemical
      Company, 54*f*
   problem of chemical wastes, 53–55
   See also Pollution

Environmental regulation, effect on
      chemical industry, 55
Enzyme blocking, role in drug design,
      85–86
Enzyme saturation, rate studies, 113
Enzyme–substrate complex
   lock and key vs. Venus flytrap models,
      84–85
   role in catalysis, 79–81
Enzymes
   all-at-once reactions, 85
   as target of medicinal chemicals, 20–22
   biological selectivity, 81–82
   biological speed, 81
   future chemical research, 89–90
   identification by biochemists, 25
   prebiotic chemistry, 31–32
   X-ray structure, 22*f*
   See also Biomimetic chemistry, Catalysts,
      Protein entries
Ethers
   as anesthetic, 78
   diethyl vs. dimethyl, 109*f*
Ethyl alcohol
   catalytic conversion to diethyl ether,
      78–80*f*
   distinguished from dimethyl ether using
      NMR spectrum, 107
   formula, 7–8*f*
Everyday life, role of chemistry, 33–45
Experimental methods in chemistry
   computational chemistry, 76
   controlled experiments, 111
   influence of computers, 70–71
Explosives, chemical vs. nuclear mechanism,
      42

F

Fabrics, synthetic, 37*f*, 54
Families of chemicals, description, 109
Families of elements, *See* Periodic table
Fats
   basic food biochemistry, 25
   soap chemistry, 2–3
Female hormones, imitation by industrial
      chemicals, 63
Fertilizers, chemistry, 40, 90*f*
Fiberglass, description and use, 46*f*
Fibers, natural and synthetic, 34, 38
Flames, chemistry, 117*f*
Flax, as natural fiber, 38

128

Human genome project, description, 27–28
Human growth hormone, role of
    biotechnology, 27
Hydrogen economy, goal of future chemical
    research, 58
Hydrogen
    benefits and dangers of use in fuel cell,
        57–58
    catalytic hydrogenation, 86–87f
    NMR structural determinations, 107
    proton catalysis, 78–81

I

Illegal drugs, chemistry, 43
Immunization, 33
Industrial chemicals
    imitation of female hormones, 63
    plastics, 35, 36f, 38, 53
    polymers, 46–47, 74, 95
Industrial chemistry
    environmental concerns, 53–55
    importance of catalysts, 77–78
    types of employment, 11f–13
    use of solid catalysts, 86–89
    See also Chemical process industries,
        Environmental chemistry
Insecticides
    history of DDT, 58–59
    purpose, 41
Insects
    chemical codes of mating, 59
    mating chemicals, 5–6
    plant repellents, 5
    use of extracts for colors, 2
Instrumentation
    use in determining chemical structures,
        105–108
    See also Computer entries, X-ray entries
Insulin, role of biotechnology, 27
Integrated computer circuit, 68f–69
Intermediates, determination of structures
    in chemical reactions, 117
Ionic bonds, description, 7
Iron, early chemistry, 3
Isomers, description, 8f
Isotopes
    role in determining reaction pathways,
        116
    use in enzyme rate studies, 85

L

Lasers, possible light sources for optical
    computers, 73–74

Law enforcement, role of chemistry, 43–44
Leather
    as chemical process industry, 12f–14
    early chemistry, 3
    role in clothing chemistry, 38
Legal aspects of chemistry
    ACS Chemistry and the Law division, 14t
    forensic chemistry, 44
    patents for chemical research, 101–102
Light
    chemical role in photosynthesis, 25–27
    optical computer design, 73–74
    self-assembling electronic circuits, 74
Limited stability of chemical products,
    environmental chemistry, 64–65
Liquid crystals, description and use in
    electronics, 74
Liquid helium and liquid nitrogen,
    superconductor chemistry, 48
Lock-and-key, as model for
    enzyme–substrate complex, 84–85

M

Magnetic materials chemistry, 69, 73
Magnetic resonance imaging, 17, 19f
Malaria, problematic control with DDT
    insecticide, 58
Mass spectrometer, determination of
    composition of chemicals, 106f–107
Materials science
    development of specialized electronics,
        67–70
    future areas of research, 46–49
    potential future advances in electronics
        and computers, 73–75
    role in modern electronics, 10
    See also Composites
Mathematics
    computational chemistry, 10, 70–71,
        75–76
    importance in chemistry, 67
    See also Computer entries
Medicinal chemicals
    difference in properties of chemical mirror
        images, 9
    general types, 20
    natural sources, 5f, 102
    pace of discovery, 10
    screening for harmful effects, 63
    See also Antibiotics, Drug entries, Health
Medicinal chemistry
    brief history, 18–20
    challenges, 23–24, 102–104

129

O

Office of Naval Research, chemical research, 43

Oil industry
as chemical process industry, 12*f*–14
use of solid acid catalysts in refineries, 88–89*f*
See also Petroleum entries

Opium, plant extract, 43

Optical computers, possible role of light for storage, 73–74

Ores, early chemistry, 3

Organ transplants, role of medicinal chemists, 23–24

Organic solvents, chemical pollutant, 55

Origins of life, prebiotic chemistry, 31–32

Overpopulation, role of chemistry in cause and prevention, 51–52

Ozone layer
beneficial absorption of UV light, 53, 61
destruction by CFCs, 60–63
hole in upper atmosphere over South Pole, 62*f*

P

Paints, chemistry of surface coatings, 47

Palladium, as catalyst for hydrogenation, 86–87*f*

Paper industry
as chemical process industry, 12*f*–14
removal of lignin from cellulose, 44–45

Pasteur, Louis
importance in chemical history, 33
view of science, 33

Patents of chemical discoveries, 101–102

Pathways, *See* Reaction pathways

Penicillin
bacterial resistance, 21
mold source of antibiotic, 21

Peptide bond breaking, role in enzyme formation, 82

Periodic table
description, 7
use to predict catalysts for hydrogenation, 87

PET, *See* Positron emission tomography

Petroleum industry
as chemical process industry, 12*f*–14
role of platinum catalysts, 87
use of solid acid catalysts, 89*f*

Petroleum
cracking, 38–39

Petroleum——*Continued*
fuels and lubricants, 38
problematic production of carbon dioxide, 65

Pharmaceutical industry
as chemical process industry, 12*f*–14
pace of discovery, 10
primary goal, 17
See also Drug entries, Medicinal chemistry

Phosphorus, chemical role in silicon computer chips, 69

Photochemistry, role in future of materials science, 73–74

Photographic film, chemistry, 45, 48, 54

Photosynthesis
basic chemistry, 25–27
challenge of imitating nature's catalysis, 91

Physicians, education in chemistry, 28

Physicists, collaborative work with chemists, 67–70, 73–75

Physics, central questions, 4

Plants
as source of medicines, 4–5*f*, 102
as source of illegal drugs, 43
dye extracts, 2

Plastic clothing, protection against chemical warfare, 42

Plastics
everyday objects, 36*f*
in home construction and furnishing, 35
in transportation, 38
recycling and degradation, 53
See also Polymers

Platinum, as catalyst for hydrogenation, 86–87*f*

Poison gases, use in chemical warfare, 42

Poisoning of a catalyst, description, 78

Pollution
automobile exhaust and smog, 56
problem of chemical waste products, 53–55
See also Catalytic converters, Environmental chemistry, Responsible Care

Polymers
description and design, 95
importance of nylon, 95
in composites, 46–47, 74
use of "sandwich" compounds as catalysts, 102

Power generation and transmission, future of superconductors, 48–49

Prebiotic chemistry, description, 31–32

Prediction in chemistry
  role of computational chemistry, 10, 15
  See also Future challenges for chemists
Prontosil, early antibacterial, 18
Protein structure
  geometry of folding, 71–73
  HIV enzyme binding, 22*f*
  polymeric, 95
  ribbon diagram, 22*f*, 72*f*
  use of NMR in solution, 115–116
  See also Computational chemistry,
    Enzyme–substrate complex, Three-
    dimensional chemical structure, X-ray
    entries
Proteins
  importance in biochemistry, 25
  role in brain chemistry, 30–31
  See also Enzyme entries, Protein structure
Proton catalysis
  conversion of alcohol to ether, 78–80*f*
  role of amino acids in enzymes, 79–81
Publishing industry, role of chemistry, 34,
    44–45
Pyrex, 35

Q

Quantum mechanics, role in computational
    chemistry, 70–71

R

Radioactive waste disposal, contributions of
    chemists, 55
Random screening, method of drug design,
    19–20
Rational drug design
  description, 19–20
  role in AIDS drugs, 23
Reaction pathways
  all-at-once enzyme reactions, 85
  analogy, 9, 72, 73, 78
  chaperone proteins, 73
  chemistry, 110–115
  future research challenges, 116–117
  role of isotopes, 116
  See also Catalysts, Energy states,
    Enzyme–substrate complex,
    Selectivity
Recreational equipment, role of chemistry,
    45
Refrigeration
  effectiveness and problems of CFCs,
    60–63
  early refrigerants, 60

Refrigeration—*Continued*
  specialized chemicals, 36
  use in food preservation, 49–50
Reproduction
  birth control chemical research, 52
  effect of industrial chemicals that mimic
    female hormones, 63
Responsible Care, Chemical Manufacturers
    Association, 53–54*f*, 64
Rhodium, role in catalytic converters, 87
Ribbon diagram, protein structure, 22*f*, 72*f*
Ribonuclease A, enzymatic purpose and
    pathway study, 113–114
Ring structures, reasons for stability, 96*f*–97
RNA
  biochemistry and molecular biology, 24,
    27–28
  cleavage by ribonuclease A, 113–114
  role in coding for enzymes, 77–78
  role in prebiotic chemistry, 31–32
  See also DNA
Robotic systems, future role in laboratory
    synthesis, 104
Rubber
  role in clothing chemistry, 38
  structure of natural vs. artificial, 94*f*–95
  vulcanization of rubber in tires, 38

S

"Sandwich" ring structures
  synthesis, 97
  use as catalysts, 97, 102
Sanitary packaging of food, 41, 49–50
Sanitation, role of chemistry, 36
Schrödinger equation, role in computational
    chemistry, 70–71
Selectivity
  enzyme–substrate interaction, 81–82
  geometry of zeolites, 88*f*–89
  importance in insecticide design, 58–59
Self assembly, in miniature electronics, 74
Self-replicating molecules, challenges for
    future chemists, 91
Semiconductor research, apparatus, 70*f*
"Shake and bake" method of
    experimentation
  as traditional method of chemistry, 6
  replacement with rational planning, 76
Shaw, George Bernard, philosophical
    quotation, 1
Side effects of chemicals
  birth defects, 63
  CFCs, 60–63

133

Thiamin, biochemistry, 82–83*f*
Three-dimensional chemical structure
    chirality, 8–9
    enzymes, 21–22*f*, 83–85
    ribbon diagram of protein folding, 72*f*
    role of computer modeling, 71
    See also Handedness, X-ray entries
Time scale, chemical reactions, 81, 117
Toxic compounds
    detection through chemical sensors,
      74–75
    waste sites, 53–54
Transistors, chemical description and
    significance for electronics, 68–69
Transportation, role of chemistry, 38–40
Tungsten, use in light bulb filaments, 68

U

Ultraviolet light
    absorption by ozone layer in upper
      atmosphere, 61
    sunscreens to prevent skin cancer, 17

V

Vacuum tube, precursor to modern
    transistor, 68
Venus flytrap, current model for
    enzyme–substrate complex, 84–85
Viruses, HIV, 22*f*–23
Vitamin B_6, synthesis, 98*f*
Vitamin B_{12}, synthesis, 99*f*
Vitamins
    biochemistry, 82–83*f*
    folic acid in human biochemistry, 19
    See also Vitamin B_6, B_{12}
Vulcanization of rubber, 38

W

Waste products
    by-product of chemical process industry,
      53
    future challenges for synthetic chemistry,
      103
    problems of disposal, 53–55
Water treatment, chemistry, 36, 41, 54
Water-based solvents, 55, 103
Weed killers, purpose, 40–41
Wood, formation of charcoal, 2
Wood burning
    comparison with charcoal burning, 2
    problematic production of carbon dioxide,
      65
    role in discovery of soap, 2–3
Wool, natural fiber, 38, 40

X

X-ray crystallography
    chemical structure determination, 81
    determination of enzyme structure,
      83–84*f*
    determination of photosynthetic center,
      26
X-ray methods
    chemical structure determination, 108*f*
    enzyme structure determinations, 21–22*f*
    limitations for structure determination,
      115–116
    use to generate three-dimensional
      structures of proteins, 72*f*

Z

Zeolites
    enzyme-based design, 90
    structure, 88*f*–89
Zinc, role in blood pressure regulation, 21